HOME OFFICE RESEARCH STUDY NO. 109

Special Security Units

by Roy Walmsley

A HOME OFFICE
RESEARCH AND PLANNING UNIT
REPORT

LONDON: HER MAJESTY'S STATIONERY OFFICE

First published 1989

ISBN 0 11 340961 3

HOME OFFICE RESEARCH STUDIES

"Home Office Research Studies" comprise reports on research undertaken in the Home Office to assist in the exercise of its administrative functions, and for the information of the judicature, the services for which the Home Secretary has responsibility (direct or indirect) and the general public.

On the last pages of this report are listed titles already published in this series, in the preceding series Studies in the Causes of Delinquency and the Treatment of Offenders, and in the series of Research and Planning Unit Papers.

Foreword

This is the first published account of the units which provide the highest level security in the prison system of England and Wales. These special security units are for those prisoners who are judged so great a security risk that they cannot be contained safely without additional security precautions even within our most secure prisons.

The report throws light on a number of features of the history and development of these units and describes in some detail the regime which operates in them. It also considers the lessons gained from the experience of the special security units that can usefully be applied in the new special units that are being introduced for prisoners who present the severest control problems in the prison system. (It is worth noting that the latter are known as 'special units' whereas the special security units are never so abbreviated.)

This report was completed at the beginning of 1988, shortly after the helicopter-assisted escape in December 1987 of two prisoners from Gartree dispersal (maximum security) prison. This escape led to a reassessment of the security requirements for the detention of Category A (maximum security) prisoners, a reassessment which includes fresh thought about the number of special security unit places that will be needed in the future. Rather than delay the report further in order to incorporate the outcome of the reassessment, it is published unchanged and thus, as the text indicates, reflects the situation at the end of 1987.

MARY TUCK
Head of Research and Planning Unit

Acknowledgements

I am grateful for all the help given me by the governors and former governors of prisons containing special security units, by other members of staff at Leicester and Parkhurst prisons and by Mr David Scagell of prison service headquarters. I also acknowledge very useful discussions with some of the unit prisoners and valuable comments from Professor Anthony Bottoms. However, responsibility for the accuracy of the information presented and for opinions expressed rests with me.

ROY WALMSLEY

Contents

Contents

Preface

At the end of 1987 there were two special security units in the prison system of England and Wales, containing just over a dozen men. These units, known as the Special Security Wing at Leicester prison and the Special Security Block at Parkhurst, are for prisoners who are judged to be so great a security risk that they cannot be contained safely without additional security precautions even within our most secure prisons. They were established in the mid-60s, along with two others at Chelmsford and Durham prisons, and in 1967 the four together contained almost 90 men. But they were much criticised and it was widely expected that they would be phased out in the early 70s when suitable alternatives would be available.*

Very little has been written about these units, about why, how, when and where they were established, how they have developed, how they are run, and why two of them have now been in existence for nearly 25 years. The opportunity to rectify this omission has resulted from the need to bring all relevant evidence to bear on the task of developing the new small units for prisoners who need special handling, not because of the exceptional *security* risks which they represent, but because of the serious *control* problems that they have persistently presented wherever they have been in the prison system. The Research and Advisory Group on the long-term prison system† thus requested a short study of the special security units which would pay particular attention to aspects, such as regimes and relationships within the units, that might assist in the process of developing forward-looking regimes in which the minimum of control problems emerged. They also recognised the desirability of having on record an account of the history and development of the special security units.

The limited time available for this study precluded a fully-fledged research project based on detailed observation and measurement of the life of the units. The decision was therefore taken to concentrate primarily on two sources: such documentation of the units as could be traced, and the

*For a sociological account of the experience of long-term imprisonment in maximum security conditions, see Cohen and Taylor's (1972) book, based on their own involvement, as teachers, in the life of the Special Security Wing at Durham. For psychological studies of the effects of long-term imprisonment in England see for example Bolton *et al* (1976), Richards (1978), and Sapsford (1978, 1983).

†The Research and Advisory Group was set up by the Home Secretary in 1984. One of its main tasks is to advise on the planning, co-ordination and evaluation of the new small units.

knowledge and experience of members of the prison service who are now, or were once, responsible for their management and administration. What follows is thus an analysis of the material gained from those sources, augmented by visits to the two surviving units and discussions with prison staff working in the units and also with some of the prisoners. Included among the appendices is some additional information, compiled from official records, that may be of some historical interest. The limitations of this study should be recognised: it does not include any data based on direct observation, nor is it based on systematic interviews or the administration of standardised research instruments to prisoners or staff. What it does aim to do is to throw light on a number of features of the history and development of these units—the most secure part of our prison system—and to describe in some detail the regime which operates in them.*

*There has never been a special security unit for women and consequently this report is only about male prisoners. See however footnotes at pp. 16 and 19.

1 Origins

"During the early morning of 12 August 1964 Charles Frederick Wilson, serving a thirty-year sentence for his part in the great train robbery, escaped from Birmingham prison. Some accomplices, the exact number is not known, scaled the wall with the aid of a rope ladder and entered the prison buildings. They located Wilson's cell after looking in several others in order to find him, released him and together returned over the wall" (Mountbatten, 1966, para 96). It was clear that under the arrangements then existing the prison system was vulnerable to similar escapes by the other seven train robbers then in custody and that special measures were therefore needed to strengthen security for such prisoners in order to thwart any further attempts.

Two days after Wilson's rescue the Home Secretary, Henry Brooke, agreed that parts of Durham, Leicester and Parkhurst prisons should be adapted for this purpose, and almost exactly a year later, on 16 August 1965, the Special Security Wings (as they were then described) at Durham and Leicester received their first prisoners, four of the remaining six train robbers in custody—Ronald Arthur Biggs having escaped from Wandsworth prison 5½ weeks previously, on 8 July 1965.

The choice of Durham as the site for a special security unit was not difficult. The prison's E wing had been used since 1961 to house men who had escaped or attempted to escape from closed prisons. Such men spent three months or more at Durham "for a period of observation under conditions of strict security and discipline" (Home Office, 1962, p 4 para 9). The wing had thus already been made specially secure and was referred to as a special unit (*op cit*) or a special wing before the work done in 1964/65. Despite the nature of its inhabitants there had been no successful escape in the period 1961–64, a fact which may of course have been influenced by the limited time that any one prisoner spent in the unit. E wing consisted of 64 cells together with two association rooms, a chapel and a library. There was a separate enclosed exercise yard. When it opened as a 'Special Security Wing' in August 1965, although it incorporated a number of additional security measures only ten cells (on the 2nd and 3rd floors) were to be used, having been specially strengthened, and it was not until after the train robbers had been moved from Durham in February 1966 that other cells were occupied. The first occupants of these ten cells were three train robbers (two of whom were in place before the arrival of other prisoners), six other robbers (including two who had already escaped from other prisons, and two who had murdered) and another murderer who had already made an escape attempt. Eight of these ten were in place by the end of August 1965; the other two arrived in September and November.

The choice of Leicester as the site for a special security unit was largely attributable to the prison's exceptionally high wall. Although only a small local prison with accommodation for around 200 prisoners (compared with 600 at Durham and Parkhurst) this feature together with the fairly small perimeter made it easier to defend. Eleven special security cells were prepared, and when the 'Special Security Wing' was opened in August 1965 it was not physically separate from the main part of the prison; however the special cells were not unlocked while other prisoners were about and special security prisoners were escorted to exercise, workshop, library etc. As a long-serving prison officer put it to me, the extra security measures had been more or less "built around" one of the lesser train robbers who remained in the wing while they were set in place. The first occupants of the unit in Leicester, in addition to the man who had been there to watch its creation, were another train robber (who was also in place before the arrival of any other prisoners) and two men with records of several previous escapes and escape attempts. All four had arrived before the end of August 1965. A third train robber arrived in September. A decision had been taken not to fill the remaining six security cells at once because the workshop was not complete by the mid-August date when the Durham and Leicester units were to open. The workshop was eventually completed in late September but before any additional prisoners had been transferred an incident in the unit at the beginning of November led to a decision to make no further transfers for the time being. However one additional prisoner was received before the end of November (he had been convicted of murdering a fellow prisoner) and another shortly before the opening of the Parkhurst unit at the end of January 1966. Incidentally, throughout this period Leicester used a cell in the unit, "temporarily" and with the consent of headquarters, for one of their 'main prison' inmates, indicating their willingness to move him as soon as the cell was needed for a 'special security' prisoner. In the end he remained until mid-August 1966, a full year after the unit had opened.

Parkhurst prison, on the Isle of Wight, was the site chosen for the other special security unit. The punishment block there, built in the mid 1930s, was entirely separate from the main prison and had an additional wall around it. There were 20 cells and an adjudication room which could be used for association. There was also a small office, bathrooms and a storeroom. A number of features were strengthened and the 'Special Security Block' as it became known was opened on 31 January 1966. All 20 cells were to be available for use, making a total of 41 at the 3 units. Eight men arrived on day 1 from various prisons raising total occupancy in the security units to 25. Four days later these were supplemented by two train robbers from

Durham SSW and another two from Leicester SSW. The eight men who arrived on the opening day included three previous escapers and one persistent escape attempter, two notorious child murderers and three large-scale robbers, at least one of whom was thought likely to be the object of a rescue attempt (another was one of the three previous escapers). The initial occupancy level of 12 at Parkhurst was only gradually increased and by the end of 1966 the unit held 16 men.

The initial selection process for the three units was as follows. Prison Department prepared its own list of prisoners who seemed the most likely candidates for the places available and then in June 1965 governors were informed of progress on the three units and asked to submit the names of any prisoners in their prison "whom you feel are such a security risk that they need containing in these maximum security blocks". They were advised that "as the accommodation is limited only really serious escape risks can be considered". A composite list was then considered at a meeting in July and decisions taken as to which prisoner would be allocated to which unit and approximately when. Initially decisions were only taken in respect of the 10 cells at Durham and the 11 at Leicester since these would be ready some six months before the 20 at Parkhurst. The original selections were not in the event the first 21 occupants of Durham and Leicester. One of the 10 Durham selections was substituted (because it was later decided that the substitute was more in need of the place) and only 5 of the 11 Leicester selections went there, because the delay occasioned first by the incomplete workshops and later by the unit incident, meant that other individuals had a greater claim on the vacancies by the time they could be filled.

Within three months of the opening of the first two units, there were serious anxieties as to whether one of them was adequate for its purpose. Following a review of security risks at the Durham unit and fears of a possible attempt at an armed rescue, an armed 'picquet' was stationed within the prison on 20 November 1965. The Home Secretary, Sir Frank Soskice, stated in a Parliamentary Answer (Hansard, 25.11.65) that he was satisfied that this was a necessary precaution and that "the picquet's instructions were to use such force only as may be necessary". It was felt that the Durham unit was considerably less secure than those at Leicester and Parkhurst and that, as soon as Parkhurst was opened, the train robbers should no longer be held at Durham. Cohen and Taylor in their book on the experience of long-term imprisonment (1972) described the arrival of the armed picquet as a "paranoid escalation" (p. 13) of the security measures but it seems that there were genuine grounds for suspecting an armed rescue attempt. The picquet was withdrawn early in February 1966 after the transfer of the train robbers.

The decision that the Durham unit was no longer adequate for prisoners who constituted the very highest degree of security risk necessitated a formal change of policy concerning its use. It was still "to function as at present and it is the intention to transfer there prisoners who are a risk to security" (memorandum to the Governor 24.1.66). Furthermore "the ten special cells should be used for those prisoners who are considered by the Governor to present unusual security risks" *(ibidem)*. But the criteria for selection for the Durham unit were to be distinguished from the criteria of those at Leicester and Parkhurst, and the level of occupancy of the wing was no longer to be limited to the 10 places in the special cells. A memorandum to governors of all male prisons (dated 20 May 1966) announced that 40 cells would henceforth be available at Durham and that the unit would be used for:

i. prisoners who, because of their continued hostility and subversive attitude in the prisons to which they have been allocated, are a disruptive influence in those prisons. This may include prisoners whose disposition to violence necessitated special allocation irrespective of security considerations; and

ii. prisoners who are an escape risk by virtue of their crime and length of sentence.

In other words Durham would be used for prisoners who presented serious control problems and those who were, theoretically at least, an escape risk. Leicester and Parkhurst on the other hand were alone to cope with the "really serious escape risks" referred to in June 1965 as those for whom all three wings were intended; at the same time their capacity was to be reduced, with only eight cells available for use at Leicester (instead of the original 11) and only eighteen at Parkhurst (originally 20). Twenty-six places were now thought to be sufficient for the prisoners who needed the very highest level of security. The distinction between the occupants of the Durham unit and those at Leicester and Parkhurst was underlined in that the memo of 20.5.66 indicated that "transfer to the Durham wing will in general be a temporary measure. Some prisoners may be there for many months but the aim of the regime will be to fit prisoners for return to more normal prison life". On the other hand "it is likely that prisoners allocated to the special wings at Leicester and Parkhurst will remain there for very long periods". It was stated that Leicester and Parkhurst would be used for:

i. long term prisoners who are not considered likely to use violence towards prison staff but who are a special security risk because of their criminal associations outside prison; and

ii. prisoners who are an escape risk and likely to enlist the support of other prisoners or outside agencies to escape.

As will be seen later, these criteria differ comparatively little from those currently used (December 1987) for the Leicester and Parkhurst units; the capacity however has been reduced further, from 26 to 17.

One other notable aspect of the origins of the special security units was the gradual emergence of the fourth unit, at Chelmsford prison. In the mid-sixties, Chelmsford was a small closed training prison for about 250 prisoners. It had not enjoyed a good security record. But following the escape of train robber Charles Wilson in October 1964, Chelmsford, like Durham, Leicester and Parkhurst was fitted with a number of additional facilities in order to ensure the safety of the prison from break-in. The prison already had what was known as a 'security wing' and this was made secure enough to hold one train robber. When the prisoner arrived on 13 July 1965, many of the new facilities were already in place and in the following two months a number of additional measures were taken, including the construction of a 'special security exercise yard' by roofing over the 'A' wing yard with steel mesh. The train robber was transferred to the new Leicester unit on 23 September 1965.

This was the first significant stage in the emergence of the Chelmsford special security unit. The second followed the escape on 22 October 1966 of the spy George Blake, who had not been placed in one of the three units but was in Wormwood Scrubs where he had been without incident, and in easy reach of the Security Service who wished to interrogate him closely, since 1961. The Blake escape led two days later to the appointment of Lord Mountbatten to conduct an inquiry "into recent prison escapes, with particular reference to that of George Blake, and to make recommendations for the improvement of prison security". But the embarrassment caused by this escape was felt to require urgent action, even before Lord Mountbatten had completed his review, which in fact he did on 21 December less than two months later. Efforts were made to identify men not at that time in the Durham, Leicester or Parkhurst units who might be security risks and to put them in the most secure location available. Two results of this initiative were first, that the occupancy level of the Durham unit rose from 16 on the day of Blake's escape to 37 by the time Mountbatten reported two months later; and second, that in the same period Chelmsford's security wing gained fifteen occupants. These fifteen were still located there four months later when Chelmsford was officially recognised as containing a 'special security wing' and was sending in regular returns of its occupants just as Durham, Leicester and Parkhurst were required to do. However, there is no reference to the Chelmsford unit in the Report of the work of the Prison Department for 1967 (Home Office, 1968), and the Radzinowicz committee (see chapter 2 para 5) which reported in March 1968 did not mention it either, referring to Durham, Leicester and Parkhurst and commenting on the fact that they contained "between 50 and 60 prisoners" (para 18). At about the time in question the three units did contain about fifty prisoners. But the Chelmsford unit flourished also and its 33 cells contained some 25–30 prisoners.

The first mention of the Chelmsford unit in a published document was in a Parliamentary Answer later in 1968 (Hansard, 24 October 1968). It was henceforth included in all references to the units, for example in the Report on the work of the Prison Department for 1968, (Home Office, 1969 chapter 2 page 8) and in the White Paper 'People in Prison' (Home Office, 1969 para 175). Thus Chelmsford emerged as the fourth special security unit: by means of what seems to have been its first official return as a 'special security wing' on 1 May 1967, it had increased the population of the units by 50% (from 56 to 84).

Such then are the origins of the special security units. The first three, containing 41 places, were created in response to the inability of the prison system in the mid-1960s to ensure the security of the prisoners who constituted the most serious escape risks, especially those with resources that might make them the object of a rescue attempt from outside. But within six months of the first prisoners being admitted, one unit had in effect been relegated to the second division in security and was to contain *inter alia* prisoners who were the most violent and subversive in the system, in other words those who constituted the most serious control problems. Only 26 places were now to be available for the top escape risks but the net for selection into the special security units had widened and their capacity was increased to 66. Following the escape of the spy George Blake more men were drawn into the three units and a fourth emerged. Within 21 months of the first two units opening, about 100 places were available in four units and 84* were filled.

*See Appendix A for total occupancy of units 1965–87; the highest number ever held in the units was 86 on 8 June 1967. See Appendix B for capacity of individual units and summary of their occupancy levels.

2 Criticisms

No sooner were the first two units opened than they attracted criticism. Publicity surrounding the extensive security measures taken raised doubts about whether the measures were excessive and conditions in the units were sufficiently humane, despite the seriousness of the crimes perpetrated by the prisoners and the importance of preventing escapes. Almost all attention centred on the train robbers and it was clear that it was to some extent influenced by the 'romance' attaching to a daring and cleverly planned robbery which involved a very large sum of money. The robbers had become if not heroes then at least 'anti-heroes'.

Home Office Ministers received correspondence criticising not only the conditions in which the prisoners were kept but also the length of the sentences which had been imposed. On the latter they could only respond that sentences were a matter for the court that tried the men and it was the duty of the prison authorities to see that the decision of the court was carried out. On the security measures and the conditions the Minister of State replied to an MP, who had written to her eight days after the first units were opened, saying that "Events have now shown that the prisoners sentenced for their part in the train robbery constitute a definite security risk and to ensure their safe custody it has been necessary to take special precautions. These measures are in no way inhumane and they will be relaxed as soon as the Home Secretary is satisfied that this can be done consistently with security" (letter from Miss Alice Bacon to Mr Dennis Hobden MP 16.9.65). But by 4 October and following a visit to Durham prison by the Minister of State, it is recorded that "both the Secretary of State and the Minister of State felt strongly that it would be wrong to keep anyone in one of the security wings at Durham, Leicester or Parkhurst for a prolonged period" (internal minute).

The stationing of the armed picquet at Durham prison led to further criticism, and the Parliamentary Answer (Hansard, 25.11.65) which was quoted in the last chapter (page 5) was in response to one of several Parliamentary Questions received. Two weeks later (Hansard, 9.12.65) the Minister of State (Miss Alice Bacon) was questioned by two MPs who were dissatisfied with the treatment of the train robbers in the Durham unit and sought reassurances that steps were being taken to improve the living conditions under which they were detained. Mr Walters MP claimed that they were being made to exercise in a yard far smaller than any other prison yard in the country, that there was an all-night light in their cells bright enough to read by and that their health was deteriorating. Mr Ian Gilmour MP commented that it was the job of the Home Office "if these very long sentences are to be imposed, to ensure that they are served under remotely tolerable conditions". They were told that special measures had had to be taken to ensure the safe custody of prisoners in the unit and that they were being treated in as humane a manner as was consistent with security. The Minister commented that some of the reports about the conditions in which the prisoners were kept had been grossly exaggerated, that the prisoners worked for about 30 hours a week in a common room and watched television on Saturdays and Sundays, and that although the exercise yard was small, it was one which had been used for many years by women prisoners at Durham. She drew attention to the building of Albany prison on the Isle of Wight "where long-term prisoners can be treated in a much better way than long-term prisoners are able to be treated at present". A week later the Home Secretary was asked specifically about restrictions on the movement and privileges of one of the train robbers in the wing. The question was full of inaccuracies but it was yet another example of public disquiet at what was going on in the Durham unit (Hansard, 16.12.65).

It was however the criticisms that Lord Mountbatten made in his report that were more authoritative and thus more significant. "The maximum security blocks which have been established at Parkhurst, Leicester and Durham", he said (para 212), "can in my view be no more than a temporary expedient. The conditions in these blocks are such as no country with a record of civilised behaviour ought to tolerate any longer than is absolutely necessary as a stop gap measure". He accepted however that "for the time being they must remain and no substantial relaxation of the security precautions can be allowed", and recommended that "a purpose-built prison is required at the earliest possible date to house those prisoners who must in no circumstances be allowed to get out, whether because of the security considerations affecting spies, or because their violent behaviour is such that members of the public or the police would be in danger of their lives if they were to get out. I call these prisoners Category A". Mountbatten, then, saw the special security units as a barely acceptable temporary expedient which would no longer be necessary once his proposed island prison (for no more than 120 'Category A' prisoners—para 214) was built.

Equally strong criticism of the units was received in March 1968 in the report of a sub-committee of the Advisory Council on the Penal System on 'The regimes for long-term prisoners in conditions of maximum security'. The sub-committee, chaired by Professor Leon Radzinowicz had begun its work in February 1967. Its most significant conclusion was that the arguments put forward in the Mountbatten report in favour of concentrating prisoners requiring the very highest degree of security in one establishment ('Vectis' Mountbatten had suggested but 'Alvington' was to be the chosen name) were outweighed by those in favour of dispersing such men amongst a number of establishments. The Radzinowicz committee recommended a strengthening of perimeter security and a liberal regime for the humane and constructive treatment of prisoners (paragraph 48), and argued that "the problem of the satisfactory containment of a small number of violent and disruptive prisoners can best be met by the establishment of small segregation units within larger prisons" (paragraph 209). But the segregation units, it will be noted, were for control purposes, security was to be dealt with by the strengthening of perimeter security in the closed prisons in which long-term prisoners were contained. The committee's comments on the special security units are in paragraphs 18 and 19 and are worth quoting in full.

> 18. In 1965 and 1966 the Prison Department established maximum security units in parts of Durham, Leicester and Parkhurst prisons. These units were established, after certain spectacular escapes, to hold long-term prisoners who were thought to be high security risks. There are at present between 50 and 60 prisoners in these units, about half of whom have previously escaped from prison during their current or previous sentences. There have been no escapes from these units.
>
> 19. We visited the three units, which are miniature prisons within prisons, and made certain proposals to the Home Secretary about improvements in the conditions. Some of these proposals were in tune with changes which had already been introduced, or were being planned, and other improvements have since been made. But no one regards the containment of prisoners in such small confined units as anything other than a temporary and most undesirable expedient. The physical limitations of the buildings preclude any major improvements in the conditions. Despite all that the staff have done and are doing, the regime of the units is unsatisfactory for men who have to be in them for long periods, and the risk of further disturbances inside the units" (see below) "is a very real one. We have therefore had very much in mind, in taking evidence and considering our report, the need to move prisoners away from these units at the earliest date consistent with security (Radzinowicz, 1968).

Mountbatten's comments are thus fully endorsed by the Radzinowicz report and improvements to the regime are not seen as removing the need for their replacement.

The reference to disturbances inside the units related principally to some troubles which occurred at the Durham unit in the early months in 1967, the repercussions of which continued throughout the rest of the year. But there were disturbances in the Leicester and Parkhurst units also. New tensions and disturbances early in 1968 came after the text of the Radzinowicz report had been finalised. The flavour of what occurred is vividly portrayed in Cohen and Taylor (1972, pp. 18–29). It centres around what they call 'the

1967 Football Mutiny' in February 1967 and 'the 1968 Chapel Barricade' in March 1968. The important point is that such unrest provided yet more fuel for the argument that such units were unacceptable as anything more than a temporary expedient. Professor Terence Morris (1968, p. 313) reviewing the Radzinowicz report for the British Journal of Criminology added his voice to the criticisms and like the report itself, made the connection between the disturbances and the conditions in the unit.

> The issue of a special prison was one on which the sub-committee found itself unable to agree with Lord Mountbatten. He said that the new prison at Alvington ought to replace the highly unsatisfactory units at Parkhurst, Leicester and Durham which had been more or less hastily prepared and which, in his view, ought not to be tolerated for a moment longer than necessary. Since the Advisory Council's report was published, there has been serious trouble in Durham as a result of just the kind of conditions with which Mountbatten was so concerned. Radzinowicz and his colleagues are not however in favour of the concentration of a large number of high security prisoners in a single institution, but rather prefer them to be dispersed in special units among three or four of the new prisons around the country.* The danger here is that these units may take on the character of the present Parkhurst/Durham type units which the Report so rightly deplores, but it is far from clear whether in deciding upon dispersal rather than concentration the sub-committee was influenced in its decision by having seen the design for the proposed prison at Alvington. By all accounts it is execrable, no more than a modern Millbank with shortcomings for the regime comparable to those of that costly white elephant of nineteenth century penology.

The Chapel Barricade in March at Durham was not the only unfavourable publicity received by the security units in 1968. Two major events threw doubt upon their security itself. At the Leicester unit some of the prisoners made what the annual Report of the work of the Prison Department for 1968 (Home Office, 1969) describes as "a determined and ingenious attempt to escape" and they "came very close to succeeding" (ch 2 para 5). In fact, several of the nine wing occupants, including the two train robbers then there, had constructed, in small sections and over a period, a ladder which was to span the gap between the top of the exercise 'cage' and the perimeter wall of the prison. It was openly in use, under the guise of racking for PE equipment, but was ready to be linked together into a ladder. Elaborate schemes had been developed for measuring the gap and had the ladder been slightly longer the plan might have worked. In short the Leicester unit was not completely secure. The perceived security deficiencies of the Durham unit have already been mentioned. And in October 1968 "three prisoners succeeded

*In fact, the Radzinowicz committee's proposal, as has been seen above, was to disperse high security prisoners amongst a number of prisons; it was not suggested that they should be held in special units within those prisons.

in getting out of the Special Security Wing at Durham, and, despite being surprised in the act of climbing the perimeter wall, one of them made good his escape'' (*ibidem*); he was not recaptured until November 1970.

It was however the events at Parkhurst in 1969 which had the greatest impact on the prison system. King and Elliott in their account of Albany prison (1977) write that ''Between August and October 1969 there were protests at a number of other prisons, especially those containing the so-called 'special wings', by the families of category A prisoners. The protests were sparked off by a new Home Office requirement that only approved persons, who had to submit photographs and obtain identity cards, could visit these men. Albany would not have been involved except that the protests led, at least in part, to one of the most serious prison riots in modern times, at nearby Parkhurst prison on 24 October 1969''. At about 7 p.m. on that day, 155 prisoners barricaded themselves in the association rooms at Parkhurst taking seven prison officers hostage. Within three quarters of an hour the barricades had been breached and the association rooms cleared of prisoners but ''in the course of the disturbance 33 prison officers and 22 inmates were injured. Nine prisoners were subsequently committed for trial on charges arising from the incident and seven were sentenced to varying terms of imprisonment as a result'' (Control Review Committee, 1984, Annex D, p. 63). What King and Elliott called ''One of the most serious prison riots in modern times'' had been provoked by the effects of a new restriction on Category A prisoners, and the protests out of which the riots emerged, had occurred particularly at the prisons containing the security units. By association at least, it was another black mark for the units.

One more significant source of criticism should be mentioned. In an article in the Times of 23 October 1970 a backbench conservative MP, subsequently a senior Government minister, wrote about ''The four security wings that are a blot on Britain's gaol system''. Pointing out that ''the conditions of the wings have been condemned by both Mountbatten and Radzinowicz'' he complained that ''almost four years after the Mountbatten report there is still no scheduled date for their closure''. He described them as an ''undoubted blot'' and commented that ''they suffer from twin drawbacks of confined space and a small population and although the numbers held in the wings have been reduced there are still 40 men imprisoned in this way—including one who has served four years in the wings''. (In fact, eight of the 39 men in the units in October 1970 had served over 4 years in the units and four of them had served over five years.) Thus, at that time, informed opinion of right and left agreed that the security units should have no future.

Reference has already been made to the strong feeling of Home Office Ministers, within seven weeks of the opening of the units, that ''it would be wrong to keep anyone in one of the security wings at Durham, Leicester or Parkhurst for a prolonged period'' (page 9 above). This feeling was reinforced by subsequent criticisms and experience, and the recognition that other solutions to the problem would have to be adopted was expressed publicly in official documents. The White Paper 'People in Prison' (Home Office, 1969), stated that, despite improvements that had been made to the conditions in the units "it remains undesirable that men should be detained for very long periods in such confined conditions" (para 175). And the annual Report of the work of the Prison Department for 1968 (Home Office, 1969) recorded the Home Secretary's acceptance of the Radzinowicz report recommendation on dispersal, announced the names of the dispersal prisons and went on: "as it gradually becomes possible to contain all Category A prisoners in dispersal prisons with a high standard of physical security, the special security wings at Durham, Leicester, Parkhurst and Chelmsford will cease to be used for their present purpose" (ch 2 para 8). It seemed certain that the security units were doomed.

3 Improvements, alternatives and closures

With ministers, respected committees, academic criminologists and public opinion apparently agreed that the special security units were acceptable only as a temporary measure, three objectives rapidly came to dominate policy: to improve conditions, to find alternatives that would render the units unnecessary, and then to close them.

Efforts to improve conditions were already under consideration before the end of 1965, and it was agreed, for example, that occupants of the units should be allowed to have wireless sets in their cells. But there seems to have been little impetus behind the good intentions and Mountbatten's strictures on the conditions were based on the situation in the last quarter of 1966. Indeed the 1966 Report on the work of the Prison Department (Home Office, 1967) makes no mention of improvements although it does include a statement of the problem: "The problem is to provide . . . at least tolerable and, so far as is possible, constructive conditions . . ." There seems at least a hint however that security considerations might limit the scope for improvements to conditions: "the needs of such prisoners and the restrictions which are necessary are given specially close attention" (page 11 para 17).

It was in 1967 that significant moves were made to improve conditions. Mountbatten's criticisms no doubt played their part and likewise the fact that persistent complaints about conditions, especially at the Durham unit, eventually erupted in February into serious disturbances there. Before the end of June Governors had received instructions as to the general conduct of the units which were "issued to help Governors improve the conditions for both staff and inmates". These instructions ran to some 47 short paragraphs in effect setting out the details of the regime and its administration. By the end of the year considerable progress has been made on a number of aspects which required building work, such as ventilation, heating, visiting rooms, bathing and shower facilities. And the Report for 1967, unlike its predecessor, stressed the negative features of the units and the need to improve conditions, and also referred to progress towards that objective (Home Office, 1968).

> The use of the special security wings to house prisoners who require the strictest security is an expedient which poses severe problems both for prisoners and for staff . . . Much was done in 1967 to improve the physical conditions in the special security wings and to provide additional privileges.

> There remained a factual basis for the many complaints about physical conditions made by prisoners housed in these wings; ventilation was inadequate and suitable work was difficult to find. Such defects however can be, and are being, remedied . . . (page 6 para 7).

As evidence of the progress of improvement the report goes on to claim that in 1967 "the special wings compared more than favourably with the general provision for prisoners elsewhere *(ibidem)*". This sounds rather like over-egging the pudding, as does the accompanying anecdote of the prisoner transferred from a security unit to ordinary conditions and complaining of privileges he was being denied. But the message comes over loud and clear. Improvements were important and they were happening.

The Radzinowicz committee's (March 1968) comments on conditions (chapter 2 para 5) testify both to the fact that some progress had been made by the beginning of 1968 and to the need for continuing improvements but the point is made that "the physical limitations of the buildings preclude any major improvements in the conditions". Efforts to make such progress as was possible dominated the quarterly 'conferences' on the security units which began in February 1968 and were held throughout 1968 and into 1969. They were attended by the Inspector General of the Prison Service, the Chief Director, the Director of Prison Administration, the governors of the four prisons, other headquarters officials responsible for their administration and the governors of Brixton, Holloway and Hull prisons*. Much time was spent in devising ways of improving the regimes including attempts to standardise procedures in the units, since inequalities between the different units had been a considerable source of dissatisfaction and friction.

Of the three objectives dominating policy for the units, that of improving conditions was the only one that involved dealing with immediate events. Finding alternatives that would render the units unnecessary and then closing them were inevitably matters for the longer-term. Nonetheless the first moves towards rendering the units unnecessary had preceded the opening of the first unit. It was intended that the new prison being built at Albany on the Isle of Wight would subsequently have a special security block for up to 100 men added to it. This block which would have replaced the special security units was already at the design stage in the first half of 1965. However, in 1966 it was decided that it should be built instead as a self-contained establishment on a nearby site, and it was just such a purpose-built security

---

*The presence of the governors of Brixton, Holloway and Hull is explained as follows: *BRIXTON*: During 1968 and 1969 a handful of high escape-risk prisoners were held in D wing Brixton along with potential unit occupants who were still unconvicted. In this period D wing Brixton was referred to as a 'special security wing' and its convicted occupants included in the total of unit prisoners. *HOLLOWAY*: Category A women were held in Holloway prison which thus constituted the nearest thing to a security unit for women prisoners. *HULL*: Plans to construct a new security unit at Hull, at an advanced stage in 1968, were subsequently overtaken by the development of the dispersal system.

prison that Lord Mountbatten advocated as successor to the units. It would be to house those prisoners that he called Category A, namely those who "must in no circumstances be allowed to get out, whether because of the security considerations affecting spies or because their violent behaviour is such that members of the public or police would be in danger of their lives if they were to get out (para 212)". The result of Mountbatten's recommendation is recorded in the Report on the work of the Prison Department for 1966 (published by the Home Office in October 1967). The units would now remain necessary "until the new special security prison at Alvington in the Isle of Wight is built" (page 11 para 17). Appendix 2 to that Report indicates that Alvington was to be, as Mountbatten had recommended, for up to 120 men.

Alvington, of course, was never built. The Radzinowicz committee recommended against concentrating Mountbatten's Category A prisoners in one establishment and in favour of dispersal (see Chapter 2) and the Home Secretary accepted this recommendation. Consequently, under the dispersal programme announced in October 1968, men in security category A were to go to one of eight prisons—Parkhurst, Wakefield or Wormwood Scrubs, and as they became available Gartree, Hull, Albany, Long Lartin and a new prison to be built in the north. Thus, in the words of the Prison Department Report for 1968 (Home Office, 1969): "As it gradually becomes possible to contain all Category A prisoners in dispersal prisons with a high standard of physical security, the special security wings at Durham, Leicester, Parkhurst and Chelmsford will cease to be used for their present purpose" (ch 2 para 8). In other words, the dispersal policy would do away with the need for the security units.

It is necessary to make clear at this point that Mountbatten's definition of his Category A prisoners, which was the definition officially adopted when categorisation was introduced in 1967, did not make category A synonymous with, and exclusive to, prisoners in the special security units. Mountbatten summarised Category A prisoners as "those whose escape would be highly dangerous to the public or police or to the security of the State". This definition thus *included* men whose escape would be extremely serious even if they were not in fact considered likely to make an escape attempt; not all such men were being held in special security units. At the same time the definition *excluded* men whose escape would not be highly dangerous but who were a source of serious disruption in prison; a number of such men were being held in the Durham or Chelmsford units, although not in those at Leicester or Parkhurst, in accordance with the criteria promulgated in January 1966 (see chapter 1 page 6).

Categorisation, in accordance with Lord Mountbatten's recommendations, meant that all male sentenced prisoners had to be labelled Category A, B, C or D, with those requiring the highest level of security being category A.

The initial process determining Category A prisoners was conducted in the following way. A governor was deputed to examine the records of over 2,000 prisoners. Just over 200 of these were selected as possibly in category A, from which a draft list containing some 140 names was prepared. At the beginning of May 1967 this list contained 138 men and 5 women* and Dr D J West (now Professor West) of the Institute of Criminology, Cambridge University was asked to conduct an examination of the prison files of the 138 men. Some of the results of this research are given in the Radzinowicz report. The initial operational list "the first list of prisoners who should be regarded as Category A" was issued as an Appendix to a Circular Instruction to Governors (No. 48/1967) on 22 June 1967 and contained the names of 139 men and 5 women. It should thus be noted that, while the Home Office had in 1966 felt that the Albany 'special security block' need only house "up to 100 men" and Mountbatten had expressed confidence that there was not likely to be any immediate necessity for a second maximum security prison so long as one was built to house up to 120 men (para 214 of his report) the first official list contained 144. When the list was issued (22.6.67) there were 83 men in the four security units, and since only 61 of these were category A prisoners†, another 78 of the first list of male category A prisoners were not being held in the units. As can be seen, it would be incorrect to assume that the first category A list was constructed to be of the size Mountbatten had expected; it must have been about 50% larger, since Mountbatten's reference to the need to build for 120 implies a belief that about 100 places would be needed at once. It would be equally wrong to suppose that the category A list would be identical to the list of prisoners in the security units; for every three category A prisoners in the units, there were another four held elsewhere in the prison system.

The dispersal programme moved on apace. By the end of 1969 Gartree and Hull were operational, making five dispersal prisons in all. The number of category A prisoners in the units had fallen to 46. In the following year there was further progress and mention of the first unit closure. The Prison Department Report for 1970 (Home Office, 1971) describes developments in paragraph 56 under the heading "Dispersal of Category A prisoners".

> Work continued on the programme to implement the policy of dispersing category A prisoners (ie those for whom escape must be made extremely difficult) from special wings to high security prisons where they can be held as part of the normal prison population. In October 1970 Albany prison on the Isle of Wight was brought into use as a dispersal prison, the sixth such establishment. The number of Category A prisoners held in the special wings at Durham, Leicester, Chelmsford and Parkhurst was further

---

*In respect of female prisoners categorisation is limited to identifying those who need to be in Category A. The remainder are not given a security rating.

†Only 7 of the 29 prisoners in the 'Special Security Wing' at Chelmsford were classified as Category A.

reduced to 40 by the end of the year. It is planned to close the special wing at Durham prison in mid-1971.

While dispersal policy was being described as the transfer of Category A prisoners from special wings to dispersal prisons, in fact less than half the original list of Category A prisoners were then in the wings, and the existence of six dispersal prisons had only reduced the number of category A prisoners in the wings from 61 to 40.

Durham, then, was to be closed in mid-1971. Its closure was first considered during the serious disturbances there in 1967 and 1968 but although the number of prisoners was drastically reduced (from 38 in mid-March 1967 to only eleven at the end of May 1968) the closure decision did not come until the beginning of 1971, with a parliamentary announcement in the House of Lords (Hansard, 17.2.71). Durham was chosen as the first unit to be closed for three main reasons: since May 1968 it had been the most under-used of the four; its closure would provide about 60 single cells for other purposes; and, because of the publicity it had received, including the disturbances, it had come to be associated in people's minds, perhaps more than any other unit, with high risk prisoners, so that its selection as the first to be closed would best serve to demonstrate the intention to abandon the wings as part of the policy of dispersal.

At the time the closure decision was made there were 40 men in the security units, 14 of them in Durham. The occasion of Durham's closure was used to plan the reduction of the population in the remaining three units at Chelmsford, Leicester and Parkhurst to 25; they were at the time regarded as having a capacity of 46 (20, 11 and 15 respectively). Nine of the Durham fourteen were to be transferred to the other units, four were to be dispersed and one was to go to a Special Hospital. A total of ten prisoners in the other three units were also to be dispersed, making fourteen in all. In the event eight of the Durham men were transferred to other units and six were dispersed. The last two prisoners vacated the unit on 24 and 25 August 1971. It had been open for just over six years—in fact for exactly 2000 days. Its closure left 24 men in the Chelmsford, Leicester and Parkhurst units.*

On the day of the public announcement of the forthcoming closure of the Durham unit (Hansard, 17 February 1971) the Minister of State, Mr Mark Carlisle visited the Chelmsford unit, noted that only seven men were occupying a wing containing 23 cells and that the Governor was confident that all seven could easily be absorbed within a dispersal prison and proposed to the Home Secretary that one of the three remaining units should go the way of Durham's. By the following January it had been decided that only 18 of the unit prisoners need remain in the units and that the 26 places in Leicester

*The unit was subsequently refurbished and since 1973 has been used to hold those women prisoners who need to be held in particularly secure conditions. At the end of 1987 Durham H wing as it is now known held three women in Category A and some 35 others.

and Parkhurst would be sufficient to allow for extra needs. The Chelmsford unit could close and the announcement was made in a Parliamentary Written Answer of 1 February 1972. The last prisoner left on 16 February, thus freeing the Chelmsford unit's 23 cells for other purposes—initially a punishment block and segregation unit. In the event, this left 17 prisoners in the two remaining units.

So, within the six month period August 1971–February 1972 two of the four units had closed, and within the twelve month period between the announcement of the Durham closure and the closure of the Chelmsford unit (Feburary 1971–February 1972) the special security unit population had fallen from 40 to 17. (It has remained below 20 ever since.) With two dispersal prisons in the pipeline (Long Lartin and Frankland) there was every sign that the intention of phasing out the units once the dispersal system was in place would be achieved.

4 New considerations and new decisions

At the end of February 1972 it would have been reasonable to predict the closure of the Leicester and Parkhurst units at least as soon as the two new dispersal prisons at Long Lartin and Frankland were fully operational. But more than fifteen years later at the end of 1987 these units were still in existence despite Long Lartin having joined the dispersal system in May 1973 and Frankland in April 1983.

Early in 1973 consideration was being given, following the recommendations of the Working Party on Dispersal and Control (Home Office, 1973), to taking over the Parkhurst unit as a special control unit such as the one subsequently established briefly at Wakefield in 1974/75. A decision had to be taken as to whether the Leicester unit could cope alone with all prisoners who presented security risks greater than the dispersal prisons could handle. It was felt that since the closure of the Durham and Chelmsford units, allocation to the two remaining units had been highly selective and that the policy of dispersing those already in them at the earliest opportunity had been continued, sometimes at the expense of adding to the control problems in the dispersal prisons. Nonetheless, it was considered that by transferring out "both those who were essentially control problems and others who do not need the very highest security, it would be possible, though not without risk of violence between factions" to accommodate the remainder in the Leicester unit. By this process the number of unit prisoners would fall from nineteen to just six. It was noted that the Leicester unit still contained eleven cells but that it "should preferably not have to hold more than eight persons in order to allow some flexibility of location and 'breathing space' in a confined situation". Seven of the thirteen who could leave special security conditions—five of whom were being held 'temporarily' in the units following the Gartree riot of November 1972—would be moved to a dispersal prison, three would need to be located in one of the new control units and the other three would be suitable for Parkhurst hospital or the therapeutic regime in the prison's C wing.

So close was the possibility of cutting out another of the units. But as the note putting together the proposals for doing so was being prepared at the beginning of March 1973, bombings took place in London and it was recognised that this development and the possibility of future similar incidents might jeopardise the proposals by producing new candidates for unit places.

Further, information was received at the same time of a plan to rescue two of the thirteen prisoners who, it was being proposed, would shortly leave special security conditions. As the writer of the note put it "these new and significant elements in the situation indicate that it would be unwise at present to work to fine margins in our arrangements for the safe containment of our highest security risks". Furthermore, the governor of Parkhurst argued strongly that the presence of a control unit in his prison would have an impact on "the generality of the population here" that would be "little short of disastrous". The role of the Parkhurst unit was left unchanged.

The reference to the London bombings is extremely important, for it was immediately clear that Irish republican bombers, once detained and convicted, would pose a serious new threat to security and there was real doubt as to whether they could be held safely within the dispersal system. This perception removed the impetus behind the policy of progressively phasing out the security units. Nevertheless, of the thirteen prisoners who had been mentioned as being suitable, albeit with some risk of control problems, for leaving security unit conditions, nine left within a year (ie by the end of March 1974) and the other four within two years; but no further consideration was given to a unit being closed.

Until July 1976, the escape of John McVicar from Durham in October 1968, when three men managed to get out of the unit but only McVicar got clear of the prison, had been the only significant failure to hold a unit prisoner. McVicar had not been recaptured until November 1970. But in the words of the Report on the work of the Prison Department for 1976 (Home Office, 1977, page 28 paragraph 106) "on 5 July 1976 three category A prisoners escaped from the Special Security Wing at Parkhurst prison by cutting their way through the wires and outer fences of the perimeter. Two were recaptured immediately and the other one within 24 hours. The escape was the subject of an enquiry by the Chief Inspector of the Prison Service. He drew particular attention to the weakness of the Parkhurst perimeter and work to strengthen it was immediately put in hand". Incidentally, it appears that the prisoner who escaped may have been at large for some 45½ hours, but he did not get beyond Parkhurst forest, adjacent to the prison.

This incident and the Chief Inspector's report convinced the department that a number of measures had to be taken to improve the security of the Parkhurst unit, over and above the strengthening of the perimeter. The seven cells on the ground floor of the unit were no longer to be used as such, the garden at the front of the unit was removed, as was the front exercise yard. A workshop was demolished and replaced by a fence. Security was generally tightened up with improved electronic equipment being introduced. This exercise necessitated the closure of the Parkhurst unit on 26 November 1976 for the alterations to take place. It was reopened on 12 April 1977. Of the seven occupants in November 1976, three were transferred to the Leicester unit and the other four were to be held by local prisons. In the event, following a suspected escape attempt one of the four was also transferred to the Leicester unit, with another Leicester inmate being transferred to a local prison. But on the reopening of Parkhurst all returned to one or other of the units.

A suspected escape plot at Leicester in early 1978 led to the conclusion that similar strengthening of security was necessary there too. The five occupants were transferred to local prisons and the unit was closed on 28 February 1978. Despite intentions that it be closed for no more than a year, the unit was not reopened for prisoners until 26 February 1980. Four of the five occupants then returned to one or other of the units from their temporary location in local prisons; the fifth had been transferred to the Parkhurst unit in June 1979.

Reference has been made to the impact of the Irish republican bombings of March 1973 on the policy of progressively closing the units. By the beginning of 1978 there were already five bombers among a total of fourteen unit occupants and it was strongly felt that nowhere but the security units could safely contain such men. Furthermore, the Leicester unit was about to close for security strengthening. Parkhurst had already been strengthened. The policy that the units would be abolished as soon as the dispersal system was fully operational no longer seemed sustainable. Seven of the eight dispersal prisons had been operating for four years or more and the addition of Frankland in a few years time was not going to change the need for the retention of the two units. Furthermore, the temporary closure of Leicester, the increased dangers of terrorism and a growth in the number of robbers with resources to arrange escapes from ordinary dispersal prisons led to the conclusion not only that two units should be permanently retained but that prudent management required serious consideration to be given to the construction of a third especially since Parkhurst was due for redevelopment when funds could be made available. It was time for a formal change of official policy.

The case for the retention of two units and the construction of a third thus rested on the arrival of the Irish republican dimension, the dangers of increased terrorism from a variety of sources and the increasing number of robbers with resources to finance their rescue. The case rested also on the perception that should one of the units be put out of action for a period, the only course of action available would be that taken in respect of the Parkhurst 1976/77 and Leicester 1978/80 closures, namely to move occupants into local prisons where they were segregated from other prisoners under Prison Rule 43 in the interests of security. This was recognised as undesirable both because rule 43 conditions were normally unsuitable for use over extensive periods and because local prisons were not nearly as secure as the special security units. What might appear to have been an alternative strategy, if one unit were out of action, to transfer such prisoners to dispersal prisons, was considered impossible, even on a temporary basis, since such establishments could not provide the close individual supervision which such prisoners required. Another alternative to the retention of two units and the creation of a third, the raising of a whole dispersal prison to the security standards of a special security unit was considered prohibitively expensive. Nor, it was felt, could such men safely be detained in a prison wing of 70 to 100 men where they could submerge themselves and avoid close observation. Thus despite the criticisms of Mountbatten and Radzinowicz, subsequent experience was seen as indicating that it was impossible to dispense with the units, and that a third was needed.

Ministers made the formal decision in mid-1978 that the special security units should continue to be part of the prison system and no longer regarded as due to be phased out. They also agreed that the number of extremely dangerous prisoners in the system rendered the existing places at Leicester and Parkhurst sufficient only if both units were in use. A third unit was accepted as necessary to avoid decanting men to local prisons and also to prevent undesirable combinations of prisoners within one unit. Shortly afterwards a bid was made to Treasury for funds to build a third unit. It would not be at the new dispersal prison then under construction at Low Newton near Durham (Frankland) because of the considerable delaying effect on the completion and opening of that prison and the financial penalties that would have been incurred on top of the basic cost of the new unit. It would therefore be part of the new dispersal prison to be built at Full Sutton on Humberside. The unit would contain 12 cells, ordinarily to be occupied by a maximum of eight people at any one time, in accordance with the policy that had developed of leaving some cells empty to allow for cells to be vacated both for essential maintenance and in case of attacks on the physical structure of the cells. Following a meeting between Home Office and Treasury officials in December 1978, Treasury approval was granted.

Thus Full Sutton dispersal prison which opened in October 1987 will contain a special security unit. This will be available before the end of 1988.* However from the early 1980s there had been doubts as to whether it would be necessary, after all, to retain three units. If it was not necessary, then once Full Sutton was available either the Leicester or the Parkhurst unit could be closed. The population of the units was remaining more or less static. In late 1978 there were eight occupants in the Parkhurst unit and five unit occupants held temporarily in local prisons during the closure of the Leicester unit, an effective total of 13. By the end of 1987 there had been little movement from this figure (see Appendix A). However between 1985 and 1987 a significant number of potential special security unit prisoners entered the system and as a result it was decided that three units would be necessary after all.

---

*See Appendix C for list of key dates in the history of the special security units.

Despite periodic complaints, usually by Leicester prisoners, about real or perceived inequalities in conditions between the two units and continued efforts to establish approximate equality (see chapter 6 for account of attempts to establish a common regime) there have been no other major developments in the life of the special security units, and certainly no major escape attempts, since the reopening of the Leicester unit in February 1980. However for the first year after that re-opening there was much tension in the unit, centring largely on inequalities between the units, and there have been other issues which have surfaced more briefly (such as militant opposition at Leicester in the summer of 1982 to searching arrangements).

5 Selection, allocation and transfer

The original (1965) selection process for the special security units was described in chapter 1 (page 5), the criterion for selection being that prisoners were "really serious escape risks". The first amendments to the criteria, made early in 1966, followed the decision that the Durham unit was no longer adequate for prisoners who constituted the very highest degree of security risk, and amounted to a broadening of the criteria to include (at the Durham unit only) prisoners who were a source of serious violence or disruption within the prison system (see chapter 1 page 6). The second criteria change, following the escape of George Blake in October 1966, drew into the units (especially into Durham and Chelmsford) additional men who were considered as possible escape risks (chapter 1 page 7). Chapter 3 showed how the population of the units was reduced under the policy of dispersal so that with the successive closure of the Durham and Chelmsford units the population which reached a peak of 86 in June 1967 had fallen by the end of February 1972 to 17. (See Appendix A for occupancy levels 1965–87.)

The selection criteria for the units thus underwent several changes in the period 1965–72. There has however been no formal change to the criteria since February 1972; nor has there been any significant change in the numbers held in the two remaining units at Leicester and Parkhurst, which have never again reached 20. In practice however, the unit population has shown some changes. For example, the seventeen prisoners in the Leicester and Parkhurst units in February 1972 once the units at Durham and Chelmsford were both closed included no Irish republican bombers, while the fourteen occupants of the units in October 1987 included seven. Thus whereas in 1972 there were seventeen 'non-terrorist' prisoners who were adjudged unsuitable for dispersal conditions on security grounds, in 1987 there were only seven. Another change to the unit population since 1972—though it probably has little or nothing to do with selection policy—is the absence now of prisoners sentenced to less than 20 years' imprisonment. The offences and length of sentence of the occupants of the units in October 1987, the occupants after the closure of the units at Durham and Chelmsford (February 1972) and also, for reference, the earliest unit occupants* (December 1965) are shown

*Excluding the man anomalously held in the Leicester unit in 1965 but not recognised as a special security unit prisoner.

in the following tables. It will be noted that seven of the 16 occupants in 1965 were sentenced to less than 20 years.

Table 1:

Offences of 1987, 1972 and 1965 unit occupants

Offences	*1987*	*1972*	*1965*
causing explosions (including explosions involving murder)	7	—	—
other murders (mainly police, child and gangland murders)	3	8	3
robbery (including armed robbery and robbery with violence)	3	8	11
other	1*a*	1*b*	2*bc*
spying	—	—	—
	14	17	16

a importing drugs
b wounding with intent
c storebreaking and larceny

Table 2:

Length of sentence of 1987, 1972 and 1965 unit occupants

Length of sentence	*1987*	*1972*	*1965*
Life imprisonment (mandatory, ie for murder)	6	8	3
Life imprisonment (non-mandatory)	2	—	—
25 years or over	4	4	5*a*
20 years or over	2*b*	1*c*	1
15 years or over	—	3	2
Under 15 years	—	1	5
	14	17	16

a Five 'train robbers' serving 30 years imprisonment each
b One sentence was reduced to 18 years on appeal
c Sentence increased to 26 years following further offences as escaper

Generally then, selection is based on the principle that the special security units are for those Category A prisoners who are judged to present so great an escape risk that they cannot be contained safely within a dispersal prison. This will be because it is considered that their escape would result in extreme danger (ie even greater danger than that posed by other Category A prisoners) to the public, the police or the security of the state, or because they have come close to success in defeating the security of a dispersal or are thought likely to command resources to do so. Thus, those eligible for admission normally fall into one of the following five categories:

a. major spies;
b. major 'terrorists' (eg offenders sentenced to long terms of imprisonment for causing explosions);

c. notorious offenders who have killed in the commission of crime and whose escape is regarded as representing a major threat to the public's faith in the ability of the State to provide adequate protection (eg police killers, some child killers);

d. Category A prisoners who have shown that they possess the technical skill to defeat standard defences or have been discovered planning an escape or rescue from a dispersal prison involving the use of firearms or explosives;

e. other Category A prisoners serving long terms of imprisonment who are thought likely to have the resources that make a rescue attempt on their behalf a significant possibility (eg those convicted of large armed robberies).

The decision that a prisoner classified as Category A shall be located in a special security unit (only Category A prisoners are now eligible) is formally taken by the Director General of the Prison Service on the advice of his senior officials; in the earlier days of the units it was taken by a Home Office minister. Such a decision will be the end result of elaborate procedures to establish whether a prisoner should be classified as Category A—a decision which itself must formally be taken by the Director General—and the further judgement that he requires special security unit conditions. In October 1987 the 14 unit occupants constituted less than 4% of all Category A prisoners*. In the ten years October 1977–October 1987 only 16 men were selected for placement in a security unit, eight of them Irish republican bombers and four of them well-resourced robbers. It is considered that all of these men could well have been the subject of a rescue attempt if they had been located in a dispersal prison.

The allocation of a newly selected unit prisoner is made by Prison Service Headquarters, based upon consideration of a number of factors. With only two units from which to choose the options are limited but the decision for Leicester or Parkhurst is influenced by the current mix in each unit (see chapter 8), perceived differences in security levels at the two units, the presence in the units of former associates of the newcomer and the current atmosphere in the units. The governors of Leicester and Parkhurst will be consulted about such allocations, since they will have the most up-to-date information on some of these aspects.

Transfers between the units occur either on the recommendation of the governor of Leicester or Parkhurst or at the instigation of Headquarters. A governor may judge that the atmosphere at his unit would benefit from such a transfer and inform Headquarters. A decision will then be taken as to

*On 31 October 1987 there were 369 confirmed Category A prisoners in the system, more than two and a half times the original (1967) Category A total of 144. Over the same period the total sentenced prison population rose by about 60%.

whether it is possible to accede to the governor's request without ignoring one of the considerations listed above as affecting initial allocations. The transfer requested may necessitate a reverse transfer—a swap. Governors indicated that experience gave them confidence in Headquarters' handling of transfers, in particular they were satisfied that if they needed to transfer out at short notice a prisoner whose presence suddenly became a threat to the stability of a unit this would be arranged. It is rarely that such a step is taken but reassurance that it can and will be carried out expeditiously is seen as important by those with the closest knowledge of running the units.

All prisoners in Category A have their status reviewed annually. In the light of advice from governors and other staff at the establishment at which they are held, a decision is taken as to whether their Category A status should continue or whether they should be downgraded to Category B. The category A prisoners in the special security units are treated exactly like all others. But in their case there is an additional option of deciding to continue their Category A status but to transfer them to normal dispersal prison conditions. A decision to downgrade would of course entail such a transfer.

Transfer to a dispersal prison is the normal means by which prisoners in the Leicester and Parkhurst units eventually leave such conditions. Most will then serve the rest of their sentence in dispersal conditions classified first as Category A and then perhaps later as Category B. In the ten years October 1977–October 1987 fifteen men left unit conditions: thirteen of them went to dispersal prisons, the other two died. Many of these men had spent several years in the units (see table below) and of the fourteen inmates in October 1987, four had spent over five years in such conditions, including two who had been so held for more than ten years.

Table 3:

Length of stay in special security units of men leaving such conditions October 1977–October 1987

Time spent in units	*Number of prisoners leaving units*
Under 2 years	2
2 years but under 5	3
5 years but under 10	9
10 years or more	1
	15

The two prisoners spending under two years in the units both died and the man who spent over ten years was one of only two unit prisoners who have spent over 15 years in such conditions. Whether or not these three men are included, the average length of time spent in the units by men discharged in the period 1977–87 was about six and a half years. Irish republican bombers tend to spend longer than the average and robbers somewhat less than the

average, in both cases because of calculations about the length of time that they continue to pose the highest security risks. The average length of stay in unit conditions has increased since 1972, because of the arrival of the Irish bombers; the average stay of those members of the February 1972 population who left unit conditions before October 1977 (15 men) was just over 5 years.

6 Regime policy and regime framework

There has never been any authoritative statement as to the objectives that should underlie the operation of the special security units. The principal objective is self-evident. Prisoners must be kept securely and must not escape. But since at least as early as 1967, it has been the policy to take such steps as are not ruled out by security considerations to lighten the atmosphere and make the specially secure confinement more tolerable. Security however must always come first.

The 1967 instructions (see chapter 3 page 15) were issued in order to improve conditions but they also amounted to a first attempt to establish a common regime for the units. By the middle of 1968 a further attempt, endorsed warmly by the Inspector General of Prisons, was being made. The Governor of Leicester prison prepared some notes for discussion of the issue including a draft timetable, draft notes for unit inmates and a draft leaflet to be sent to relatives. Throughout this period regime details were adjusted to remove variations which remained or newly emerged as sources of conflict. Late in 1973 a third attempt was made to codify a common regime. A summary was made of all existing instructions (some 68 were listed) and the governors of Leicester and Parkhurst, by now the only prisons with security units, indicated which of the instructions were in force in their units. A meeting was held at which conclusions were reached as to the scale of existing differences and appropriate solutions. This 1973 attempt had been inspired by problems over the question of payment for food out of private money. The meeting ended with three main differences unresolved, namely association, earnings and, unfortunately, food purchases from private cash. It was recorded that once these points were resolved an agreed note would be issued, and in April 1975 an amended version of the 68 instructions, now reduced to 47, was sent to the two governors. The major changes, which had been agreed at the 1973 meeting, were the deletion of all references in existing instructions to work (which was now optional—see chapter 7) and to treatment (these had envisaged a planned programme for each prisoner which had been found "quite unrealistic however desirable in theory"). Despite this third attempt at establishing a common regime, problems due to variations between the units, though at their greatest in the earlier days of the units, have rumbled on over the years, occasionally (as in the year after the re-opening of the Leicester unit in 1980) causing considerable tension between inmates and staff, and they are still a periodic source of grievance today.

Attempts to establish a common regime have concentrated on the details of day-to-day living in the units—pay, food, letters, visits, work, recreation, hobbies. They have not addressed the more central questions of the overall objectives of the units and this seems to be because the objectives are not seen as differing from those of the rest of the prison system. These are formulated in the most recent statement of the task of the prison service (Annex A to Home Office, 1985) and may be summarised as:

a. to keep all sentenced prisoners in custody, with such degree of security as is appropriate, having regard to the nature of the individual prisoner and his offence;

b. to provide for prisoners as full a life as is consistent with the facts of custody, in particular making available the physical necessities of life; care of physical and mental health; advice and help with personal problems; work, education, training, physical exercise and recreation; and opportunities to practise their religion;

c. to enable prisoners to retain links with the community and where possible to assist them to prepare for their return to it.

In conjunction with these overall objectives Leicester prison produced in 1985 a set of special objectives for its Special Security Wing. In part they simply re-echo the overall prison service objectives but they also include implicit recognition of three additional aspects. The objectives of the Leicester unit are stated as:

a. to hold in special security conditions certain high security risk prisoners as may be directed by Headquarters;

b. to control the prisoners in the Special Security Wing by means of the activities, routines and procedures laid down for the Wing;

c. to ensure the physical and mental health of prisoners by the use of the specialist and physical resources available;

d. to ensure the safety of prisoners by being aware of the factions and groupings which may occur from time to time and by taking any necessary action to avoid violence between prisoners;

e. to ensure the health and safety of staff. This will be attained by constant vigilance, mutual support, the use of the available liaison and support staff of all levels, and adherence to the laid down procedures. As the dynamic of the Wing is subject to change, suggestions for the improvement of the procedures will always be considered.

The three additional aspects come in points b, d, and e. First, it is recognised that control of the prisoners is effected not by rigid disciplinary requirements but by the regime's activities, routines and procedures. Second, it is recognised that the safety of prisoners will depend, at least in part, on staff awareness of "the factions and groupings which may occur from time to time." In a small community, closely confined, this is undoubtedly of central importance. Lastly, emphasis is placed on the need for staff support and on the changing dynamic of the Wing. All these points are considered more fully in chapter 8.

Consideration of regime policy leads on naturally to consideration of regime framework. The framework of the regime, in other words the buildings and space, manning levels, management practice and time unlocked are, like everything else to do with the units, dependent on considerations of security. That is not to say that only one type of building would satisfy the demands of security or that a particular manning level, management style or timetable is essential. But whenever decisions in these areas are being taken it is security that has to be the prime consideration. This is inevitably the theme that runs through the following description of the framework in the two units.

The Leicester unit was constructed on the ground floor of the prison in what was originally the end of one galleried wing. The unit was rendered largely self-contained by the construction of walls separating it from the rest of the ground floor of the wing and a ceiling separating it from the higher floors/ landings. Access is through doors with electronic locks which are operated by remote control. There are now only eight cells available for occupation by prisoners (one of them a special punishment/time out cell), the other three having been taken over by a handicraft room, a small gymnasium and a kitchen; there is also a shower and toilet area and a room which is variously used as a TV room, a dining room, an association room, a classroom and an adjudications room. The unit also has its own visits room and exercise yard and the staff office, with a large window which allows full observation of the unit, is situated between the unit and the rest of the wing. When not in a cell, a room or the exercise yard prisoners have access to the space from the unit office to the end of the wing between the cells/rooms situated on either side; this is an area of some 25 metres × 6 (150 square metres), in which are located a snooker table, a table tennis table, some exercise equipment and four chairs for the prison officers on duty in the unit. The exercise area, which resembles a cage in which one current unit member refuses to set foot for that reason, measures about nine metres by eight (approximately 75 square metres).

There is no doubt that the physical conditions are not very satisfactory for long-term inmates. There is little natural light in the central recreation area and the ceiling is low, giving a somewhat cramped feeling to the unit, which over a long period could become claustrophobic. Apparently some staff do find the unit claustrophobic, while others do not. It was variously described to me as "like a goldfish bowl" and "like one large cell" and both the Leicester and Parkhurst units have been described as 'electronic tombs'. All these descriptions seemed somewhat too negative to a short-term visitor at a time when the unit atmosphere in terms of relationships was clearly good, but they would clearly be apposite, from the viewpoint of a long-term occupant, especially if the atmosphere was less relaxed. In comparison with the Parkhurst unit, that at Leicester has fewer facilities and considerably less space. The view from the unit is also very limited: prison buildings and the exceptionally high prison wall.

The Parkhurst unit is housed in the old punishment block and is entirely separate from the main prison accommodation. It is a two-storey cellular block and extra security measures installed, in addition to electronic surveillance and electronic control of access/egress, include a new wall around the block and, following the temporary escape in 1976, an additional fence replacing one of the exercise yards and gardens and also the workshop. There are now only 13 cells available for occupation and the certified normal maximum number of prisoners is 10. These are all on the upper floor. The ground floor cells, now used mainly for storage space, have not been used as inmate accommodation since 1976. The unit also has a small kitchen area, a large association room which is also used as a dining room and a TV room, a large workshop-type room for hobbies and art and a large gymnasium. There is also an additional black and white television in a former cell. The unit shares a visiting room with high risk prisoners in the main (dispersal) prison but is particularly well-equipped for outside exercise space. There is a large compound area (measuring some 95 × 35 metres) with a concrete path, used regularly as a running track, around the perimeter. The area is divided into three parts, with the first containing a tennis court (rarely used for that purpose), the second a lawn and the third a garden with a small greenhouse in which the prisoners may grow their own produce. As a governor with experience of running the Parkhurst unit has put it, the compound area "is particularly valuable in providing a large space where prisoners can get away from the enclosed atmosphere of the unit, get fresh air and sunshine and, perhaps more important, get away from each other".

The conditions at Parkhurst are clearly more suitable for long-term inmates than those at Leicester. The general atmosphere is of a light airy unit where space is not a problem. At any one time there is a variety of places where prisoners may choose to be. This allows some privacy and, despite restrictions on the use of the compound, enables prisoners who are irritated with each other to avoid prolonged contact. The garden is seen as having immense therapeutic value. Not only does it provide a place outdoors where prisoners can forget their immediate environment but it is also a place where they can produce and do something worthwhile and, perhaps most important, make plans for next year. It is a mechanism for helping prisoners cope with time. Access to the compound is nonetheless limited and unpredictable (see chapter 7) and it would be a mistake to overstate the positive aspects of conditions at the Parkhurst unit. The demands of security, and especially the continual surveillance by staff and cameras can make this unit too seem claustrophobic. As is to be expected, there is sometimes unease in relations between certain prisoners, and the negative effects of such tension, from the point of view of everyday living, are inevitably exacerbated in so closely confined a community.

Thus, in both units, the nature of the buildings and the limited space, and other security requirements such as continual surveillance, restrictions on movement and the sparseness and comparatively unchanging nature of the human contact available, make them a distinctly unattractive environment in which to spend long periods of imprisonment.

As has been made clear, the continual surveillance is achieved both electronically and by the efforts of staff. And one of the most distinctive features of the regime framework in the units is the high manning levels that are maintained. In the Leicester unit, where the normal inmate capacity is currently fixed at seven and the population since 1980 has usually been six or seven, the normal staffing level during the main part of the day is fixed so that the number of basic grade officers on duty exceeds the number of prisoners by two; a senior officer and a principal officer are also on duty at the same time. If occupancy falls below seven prisoners then the number of basic grade officers exceeds the number of prisoners by three and if it falls below five prisoners the excess is four. In the Parkhurst unit where the normal inmate capacity is currently fixed at ten and where the population since 1980 has also remained at around six or seven the staffing level during the main part of the day is fixed at ten basic grade officers plus one senior officer and one principal officer.

Manning levels at both units, which are based, I understand, on a system outlined in a 1977 Manpower report, are extremely expensive to maintain. The obvious question is whether the presence of staff in such large numbers is really necessary. Opinion seems to be divided on this matter. On the one hand it is pointed out that at the Leicester unit only four of the basic grade staff 'on duty' are actually in the wing at any one time and that the rest are 'standing down' unoccupied in the wing office (they change over at hourly intervals); similarly at the Parkhurst unit it is claimed that it cannot be justified having ten basic grade officers supervising about six men with all of them locked up in escape-proof conditions. On the other hand it is argued that the 'standing down' officers at Leicester are on hand in case of trouble and that the lay-out at Parkhurst, with more rooms to supervise, requires more staff 'on the floor' than are needed at Leicester.

A balanced view on this issue must take account of a number of factors in addition to those already mentioned. These may be represented by the following statements and opinions, gleaned from documents and from discussions with governors; the first five argue for high manning levels while the other five suggest that lower levels would be acceptable:

a. most governors with experience of managing the wings maintain that it is desirable to have at least one more member of staff available than there are prisoners. The reason for this is the feeling that such a staff presence will inhibit assaults by inmates both against staff and against other inmates.

b. any significant reduction in manning levels would be a message to inmates of weakness or vulnerability. If staff behave in a low-key fashion, their presence in numbers need not raise tensions.

c. a high staffing level is needed not because security unit inmates will be violent if not closely supervised but to deter the possibility of trouble breaking out at any time.

d. when dealing with notorious prisoners it is better to be over-manned than under-manned. If trouble breaks out the fuss would be enormous and the response that would be necessary would inevitably be very costly.

e. a high proportion of security unit staff are in their 50s. If all were young and supremely fit the staffing ratio could be lower. But it would not be wise to replace them all with younger but comparatively inexperienced officers.

f. high staffing levels derive from the time when the units were used not only for security risks but also for control problem prisoners. But now that selection is confined to men who are primarily a security risk, few control problems arise in the units and the justification for high levels of staffing has gone.

g. one of the dangers of staffing the security units is that the generally relaxed atmosphere and easy relations between staff and inmates (see chapter 8) can 'condition' staff into being less vigilant on security matters. High staffing levels reduce the sense of responsibility that any one member of staff will feel and make him more likely to be bored with the work. Indeed it is claimed that the serious escape attempts at Leicester (1968) and Parkhurst (1976) both owed much of their near success to the conditioning of staff.

h. although there may be an underlying fear that high-risk prisoners gathered together would cause control problems, this has not proved to be true and change has been delayed only by the inertia of local management over the years, together with the fact that manpower teams have been deflected from the task by the mystique surrounding the security units of which they had little knowledge.

i. there is no evidence to suggest that staff need to be present in greater numbers than the prisoners in the interests of their own safety. The ethics of the units are such as to make attacks on staff highly unlikely.

j. although staff considerations differ in the two units because of the differing designs and physical security arrangements, there may be no need for a discipline staff presence in every part of a unit whilst prisoners are present. It may be perfectly adequate for staff to have a patrolling rather than a static function.

These statements and opinions need to be considered in conjunction with the arguments and counter-arguments quoted in the preceding paragraph. They do not point unmistakeably to one conclusion, but perhaps three comments are appropriate. First, both the nature of the inmate population and the low record of control problems in the security units argue for some reduction in present staffing levels, so long as the physical security of the units is adequate to its task. Second, arguments of economy and of staff motivation and efficiency also point towards some reduction in levels. Third, any reductions that take place should be undertaken with great caution.

A third aspect of the regime framework is the general style of management practice adopted. This inevitably helps to create the regime atmosphere, it affects the freedom or otherwise of inmates to participate in the regime or at least make suggestions that will be seriously considered and it demonstrates how far the governor is willing to go in making life for the inmates as tolerable as possible within the constraints of security.

What seems absolutely clear and accepted by governors, staff and prisoners alike is that management must be strong, with a clear idea of what is wanted. It must be unobtrusive with a highly professional approach and it must be well-controlled, creating no unnecessary aggravation for staff or prisoners. Happily there seemed general agreement that this policy is followed now in both units and has been followed in recent years by successive governors. It was important that everyone should know where they stood and what was expected of them. The regimes of the two units had not changed much over the last 15 years and routine was important. Informal access to the governor was much valued by the prisoners in one unit; they considered indispensable the opportunity of bringing to his attention from time to time matters of concern which did not, at least at that moment, require a formal interview with him and a formal decision. Everyone recognised that security was all important, the prime objective of the units, but the overwhelming feeling was that within the constraints of security requirements conditions should be as tolerable as possible, for the benefit of staff and prisoners alike. Both groups had to exist for long periods in close proximity and close confinement.

The decision to manage the units in such a way as to make conditions as tolerable as possible means that efforts are made to see that relations between staff and prisoners (see chapter 8) are good and that the atmosphere is relaxed. By all accounts this is generally achieved, although there are periodic grievances that have to be resolved and, inevitably, misunderstandings and personal characteristics and relationships create occasional tensions. One of the ways in which good relations are maintained is by allowing prisoners a degree of self-determination. Clearly, this is not on any major level but in domestic details such as the allocation of chores and the choice of visiting days they are allowed some day-to-day control over their own lives. There are also long-established arrangements whereby they may supplement their prison pay by some use of private cash and thus augment the regime in certain closely regulated ways. Efforts are often made by inmates to persuade successive governors that a variety of concessions should be allowed and governors regard it as important to give serious consideration to such requests and sometimes to grant them, where there is no good reason for them to be rejected on security or other grounds. Again, where a particular request cannot be granted it may be that an analogous amendment to existing practice can be made instead. And when, on grounds of economy, prisoners in one unit had to be locked up half an hour earlier each night, they were able to negotiate, as a sort of *quid pro quo*, compensating concessions. Such flexibility and willingness to meet prisoners half-way undoubtedly contributes to the normally relaxed atmosphere in both units. But where a request bears on security it is invariably refused, as when recently prisoners in one unit protested at the introduction of an element of unpredictability in respect of the time of their exercise periods. Nor would compromise be allowed in respect of the rule whereby prisoners are at irregular intervals transferred between cells. This is a security precaution and though unpopular—prisoners cannot feel that a particular cell is ever theirs—there is no room for negotiation here. But while rules cannot be broken and nothing detracting from security is in the least acceptable, there is room for some concessions within the rules and it is here the governors use their discretion and generally try to be as amenable and conciliatory as possible.

It will be apparent that governors not infrequently find themselves having to make difficult decisions about the extent to which they are prepared to improve the existence of prisoners, many of whom have committed the most appalling crimes. This dilemma has been evident since the early days of the units, and is common, in differing degrees, to all prison governors whatever the nature of their prison. It is a dilemma which is also recognised by senior prison officials since they must be prepared to offer guidance, and sometimes instructions, to governors as to what is and what is not acceptable. The considerations which have led to general agreement that current practice in the running of the security units is correct are conveniently summarised in two unpublished papers, the former by a senior prison official writing in 1971, the latter by a senior governor writing at the beginning of 1987.

After describing conditions in the units in 1971 the official writes:

> These conditions compared with those outside Special Wings could be said to amount to "feather-bedding". There is however, another side to the story. The conditions have evolved as part of the price of peaceful co-existence. Recent protests in the wings have been confined to passive hunger

strikes and the occasional individual outburst. The violence of 1967–68 has not been repeated and it would not have been acceptable to the public or to the staff if it had been. Fears over hostage-taking have disappeared. Additionally, however, staff complements have been increased. Staff are better trained and prisoners are often out-numbered by the staff. Observation of prisoners by staff is much greater than elsewhere. The population in the wings does not vary much so that the restrictions on space which undoubtedly exist combined with adequate staff and few changes of face, all contribute to psychological pressures. Only at Chelmsford is there a football field of a size equivalent to that in other dispersal prisons. Although the exercise area at Parkhurst has been enlarged in the last two years, where there is also a garden, the overall area is still small. At Leicester the exercise area is no larger than two tennis courts and the only other open space available is the wing 'corridor' itself.

To sum up, in 1965–66 conditions in the wings were not acceptable and were criticised. In 1967–68 there was a succession of acts of violence in the wings, and our reaction was to improve conditions as a counter to the violence and to the original criticism. At the same time, the wing populations were steadily reduced. Had we not made conditions easier, criticism would have continued and violence directed towards staff would have escalated. Solitary confinement on the continental pattern could have been adopted, but would again have been unacceptable.

The senior governor writing at the beginning of 1987 records that in 1970–71 he served as an assistant governor with responsibility for one of the units and was able to observe at first hand the change from the restrictive regimes which operated in the 1960s and led to the troubles of 1967–68 to the management pattern which has since 1970–71 continued unchanged and is now accepted as the most effective way of managing the security units. The change, he says, "consisted, according to one's point of view, in the granting of sensible regime concessions in an attempt to compensate the prisoners for the confined existence that security considerations demanded, or the erosion of staff control to the point at which prisoners ruled the roost." And in looking at the regime in today's units, he comments: "As regards the basic routine of daily life . . . one is tempted to wonder if maximum association and the absence of work other than of a domestic nature are appropriate to men whose very notoriety and the seriousness of whose offences would seem to disqualify them from such privileged treatment. One must, however, I feel, balance one's instinctive feeling that badness should not be rewarded against certain self-evident facts. These are:

a. That the . . . [units] . . . are dedicated to security and not to control. The security requirements alone constitute substantial deprivation in the form of restrictions on movement and the sparseness and unchanging nature of the human contact available to the prisoners. To further restrict these men would add nothing to the degree of security achieved by the physical structure and, since no control problems of any significance have arisen, could only be seen as unnecessarily punitive.

b. That the attempt to provide meaningful work which prisoners in the units were compelled to do foundered many years ago*. I see nothing in the present industrial scene in prisons to persuade one that anything better could be achieved now.

c. That to attempt to impose restrictions on unlocking without a reason understandable to both prisoners *and staff* . . ., or to impose a work environment where none has existed in the memory of anyone now forming part of either group, would meet resistance, the overcoming of which, even if it were achieved, would have little point."

This then is the *modus vivendi* that has become established in the special security units. As another governor has put it: "In theory running a . . . [security unit] . . . is an almost impossible task. The men contained are all dangerous, all without hope for the future and individually are unpunishable and undeterrable. Men of differing ages and philosophy have to live in close contact with each other with few other contacts for very long periods of time in conditions of maximum security and constant observation by staff." If the units can be made to work, but experience shows that this depends on operating a regime that is more liberal than many would instinctively favour, this surely is how they must be managed.

One further important piece of the framework of any prison regime is the time that prisoners spend 'unlocked' (ie out of their cells or free to leave their cells). The restrictiveness of a regime is sometimes judged by this criterion with especially critical attention drawn to prisons in which the inmates have to remain locked up for as much as 23 hours a day.

The position at both the Leicester and the Parkhurst units is considerably more favourable than this. Indeed prisoners are unlocked for eleven hours a day. The weekday locking and unlocking times at the two units are approximately as follows:

Leicester	*Parkhurst*
7.50 a.m. Cells unlocked	8.10 a.m. Cells unlocked
12.00–1.30 p.m. Cells locked	12.00–1.10 p.m. Cells locked 5.00–5.45 p.m. Cells locked
8.30 p.m. Cells locked for night	9.00 p.m. Cells locked for night
Total Time Unlocked 11 hours 10 mins	*Total Time Unlocked* 10 hours 55 mins

Only two of the eight dispersal prisons (Parkhurst main prison and Long Lartin) match the security units in time unlocked. In the other six, prisoners are unlocked for between 8½ and 9½ hours a day.

*See chapter 7 and appendix D. Compulsory work at the security units was discontinued early in 1972.

Prisoners in the security units are thus locked in their cells for about two hours a day less than those in most dispersal prisons. As will be seen in chapter 7 there are other aspects too (eg visiting arrangements) in which unit prisoners are treated more generously than other Category A men. But in general the privileges allowed to unit men and other dispersal prisoners are the same, except that the small size of the units and the limited variety of company reduces possibilities. It is said that a lot of unit prisoners would opt to remain in a unit rather than be transferred to a dispersal and that they see the unit privileges and concessions as generous, but at the same time it is pointed out that most unit prisoners have never been in the dispersal system and that if they went there they would realise how restricted the opportunities in the units really are.

The principal plank of policy for the security units is that prisoners shall be held securely, despite their real potential to escape or be rescued. The secondary objective is that, so far as is consonant with security, conditions will be kept as tolerable as possible—with prisoners compensated for the necessity to keep them confined under close scrutiny in a small community with limited opportunities by efforts to ensure, in the interests both of avoiding control problems and providing staff with a bearable atmosphere in which to work, that the regime is relaxed and civilised and provides the prisoners with opportunity for a limited amount of control over at least some areas of their daily lives. But there remain considerable fears, most of all amongst the prisoners themselves, that long-term confinement in these conditions may be seriously damaging (see for example Cohen and Taylor, 1972, pp. 104–111). An often repeated criticism is that the units merely contain the prisoners and that it is difficult in such an environment to keep the brain alert. Difficult it certainly is but it is not impossible. Cohen and Taylor referred to the efforts of the prisoners in the Durham unit in the 1960s to keep mind and body active. The same occurs today, and probably the most determined of current inmates takes care to fill each day with almost every minute accounted for, writing, reading, taking vigorous exercise and pursuing various hobbies. As one officer put it, "it could be soul-destroying here after a number of years but it's up to the individual". The general view however seems to be that, while conditions in the units are as one governor put it "not particularly awful and absolutely necessary", few can sustain so demanding a self-imposed programme as the man mentioned above and that, as another governor put it, "the best that can be hoped for is that they [men who stay nine or ten or more years in the units] will become institutionalised". Individuals were quoted to whom this had apparently happened.

7 Regime activities and contact with the outside

In any prison, and especially a prison for those undergoing long sentences, it is important that there should be opportunities for inmates to engage in a number of activities and so occupy themselves during the day. This is necessary as much because idleness provides scope for trouble-making as because a humane prison system requires that a constructive programme should be provided for those imprisoned. The Control Review Committee (1984) was very clear that "from the point of view of managing prisoners' sentences, the need is to ensure a wide range of facilities (sc. for activities) at each security level" (para 93) and Dunbar (1986) argues that activity can enhance the security and safety of the establishment. Prison activities are thus seen as significant in respect of both security and control. Likewise it is important that prisoners, almost all of whom will eventually return to live in the community outside, should be enabled during their sentence to maintain good contacts with family, friends and others who are close to them, since such contacts make the prison sentence more tolerable and the breakdown of supportive relationships seems likely to reduce the chances of their settling down as worthy citizens in the future.

One activity which features on the daily timetable of most prisons is work. Prisoners are required to undertake employment, for which they receive a nominal amount of pay. Unfortunately it has long been impossible to provide a constructive eight-hour working day in English prisons, let alone one involving work similar to that which the prisoner may reasonably expect to be undertaking after his release. The average prison in England and Wales now offers work for only 20 hours a week and this usually to fewer than half its prisoners. In the special security units compulsory work other than cleaning duties ('domestic chores') was discontinued in 1972. It was always found difficult to provide suitable work and adequate workshop provision in unit conditions and there were persistent control difficulties, with the requirement to work generating a general lack of co-operation and being a source of considerable friction.*

There has been no significant pressure for its re-introduction. Even occasional failure to undertake the domestic duties is unlikely to be a source of punishment. Pride in living in a clean environment and normal community

*See Appendix D for a brief history of work in the security units 1965–72.

pressures ensure that the duties are done and the units are undoubtedly extremely clean.

All unit prisoners receive a common rate of pay but because they do no other work than cleaning the sum they receive is lower than that of dispersal prisoners (£2.87 per week at 31.12.87, compared with average dispersal prison pay of about £4 per week). They are additionally entitled to spend from their private cash a sum of £115 per year plus a further £113 per year for hobbies (rates at 31.12.87). In addition to these standard allowances, shared by all convicted prisoners, Leicester allows a small sum from a local welfare fund to be used for hobbies expenditure and Parkhurst allows an extra sum from private cash for the purchase of additional foodstuffs.

In the absence of compulsory employment to fill the day the main activities are hobbies, education and exercise and recreation. Reference has already been made to the 'hobbies allowance' and the opportunity to supplement this from private cash. Hobbies are very much recognised as an important means of enabling a prisoner to spend his time in a security unit constructively. Hobbies have always been part of the unit regimes, with rug-making, soft-toy making and minor carpentry work being undertaken in the early days of the units, although originally as part of compulsory employment. Art and woodwork are other long-standing hobbies. In the Leicester unit there is now a hobbies room (once a cell) for woodwork, carving and model-making. Soft toys are also made. One prisoner has a talent for painting and sculpture and took 'A' level Art examinations in summer 1987. Leathercraft and woodwork require a variety of tools which are a potential risk in a unit where security is the prime concern, but their use is carefully controlled and they are kept, all 80–90 of them, on a shadowboard which enables staff easily to monitor their presence. At the Parkhurst unit similar hobbies are pursued and a former workshop area is used for carrying them out; this is the only area of the unit where hobbies other than painting may be pursued. Tools are always checked when prisoners leave the workshop and they have to be accounted for at all times and kept in a locked steel cabinet when not in use. Under no circumstances can prisoners take tools from the workshop area. At both units, in accordance with Standing Orders the Deputy Governor's permission must be obtained before any prisoner is allowed to undertake a new hobby.

An important activity at both units, but especially at Leicester, is cooking. Indeed in that unit it is probably the most important 'hobby' from the prisoners' point of view. Over the years various practices have been adopted in the units with regard to the provision of meals, with them sometimes being brought in from the main prison kitchen and at others being brought in as 'dry rations' for cooking by the unit prisoners. At Leicester the kitchen, a former cell, has become a focal point of the prisoners' social activity, a 'social catalyst' as one inmate described it. An officer collects rations from the main kitchen and after careful checking they are cooked in the unit by inmates, who decide among themselves who is to be cook. They may within the prescribed limits supplement the daily rations from private cash and earnings. The facility is much valued by the Leicester inmates who said that it was used for 21 man hours every day; however its very popularity has led to it being a considerable drain on their allowances and they claimed that this year's allowance would be exhausted within 9 months. At Parkhurst too the unit has a small kitchen area. Whilst all meals are at present provided from the prison kitchen, prisoners may purchase food from earnings and private cash and may cook this in the kitchen as a supplement or alternative to the prison diet. Food may also be cooked that has been grown in the allotment area of the exercise compound. The difference between the two units stems from the time when the unit 'cooks' at Parkhurst refused to prepare meals for an unpopular fellow prisoner. The view taken by staff was that they should cook for all or none. Incoming food at Parkhurst is searched and the dangers of messages or contraband being passed in are regarded as slight. It was however to avoid such contact that self-catering was originally introduced, and one governor suggested that if a prisoner was unpopular enough to be denied meals cooked by his fellows, the provision for him alone of the somewhat less appetising prison meals brought in from outside might have the effect of persuading him to make greater efforts to gain acceptability.

Education facilities available to inmates in the security units are limited. They have access to library services of which many make considerable use. They can undertake correspondence courses and they can attend classes. All education has to be demand-led and inevitably classes are small and occasionally operate on an individual basis. But education, like hobbies, is encouraged in order to preserve and develop prisoners' knowledge and skills and in order to occupy their minds constructively. The average level of intelligence of men held in the units is generally agreed to be higher than that of prisoners in general and some are highly intelligent. Cohen and Taylor noticed this in the Durham unit twenty years ago; today's more focused selection policy may have accentuated the distinction. Current education courses at both units include art, leatherwork and soft toy making and there are also classes at Leicester and at Parkhurst in Spanish. There is currently a request for courses in the Irish language from some IRA prisoners, and Open University courses are being considered. The selection of teachers requires particular attention in a security unit. They must be acceptable on security grounds and they also have to be acceptable as individuals to unit prisoners and staff.

Exercise opportunities are available in both units. At Leicester one hour's exercise is allowed in the morning and one hour in the afternoon, increased by half an hour each session during the summer months. As has already been said, the exercise yard, although of reasonable size, has the appearance of a cage with a heavy anti-helicopter device fitted over the top. Exercise is optional and one inmate currently refuses to use the yard. Physical education equipment is also available: a running machine, a rowing machine, a sit-up bench and an exercise cycle. The equipment is at present used by three of the unit's seven prisoners. At Parkhurst, the compound affords considerably more generous exercise space and is understandably more popular. It provides opportunity for tennis, running, walking and working in the garden. In practice the main activities are jogging and walking. The formerly very generous exercise periods have now been reduced on security grounds but prisoners can still spend at least two hours a day on exercise. Physical education material at Parkhurst, located in the gymnasium, includes weight-lifting equipment and a punch-bag, but is only used by one man at present.

Other recreational activities not involving exercise are naturally popular. At Leicester the association/dining/television room also contains a video. There is a weekly feature film. The table-tennis table is rarely used except as a surface for arts and crafts (eg painting and soft-toy making) but the snooker table is much in demand with almost all prisoners playing. Staff used to join in but this was discontinued when the need to complete a frame in which a staff member was participating was found to be delaying meal-breaks for other staff. At Parkhurst too the association/dining/television room contains a video on which hired films are shown. A second television room—a cell—contains a black and white set. Again, paralleling the situation at Leicester, there is a snooker table but the table-tennis table, in the craft room, is now used for other activities.

To sum up, when prisoners are out of cell they may at Leicester use the kitchen, the association/dining/television room, the PE equipment, the snooker table, the hobbies room and also the shower and bath. At Parkhurst the position is the same except that there is a second TV room. Outside exercise is available for at least two hours a day. Additionally, classes are arranged at both units as necessary and both units are visited regularly by the prison chaplains who will arrange religious services in the units if so requested. Other regular visitors are a welfare officer, a prison officer PE instructor, the librarian, and, every day, a medical officer and the governor. Formal applications from prisoners are taken each morning by the principal officer in charge of the unit. A typical daily timetable at the Leicester unit would be:

7.50	Cells unlocked
8.30	Breakfast
9.00	Applications
10.00–11.00	Exercise
11.05–12.00	Lunch
12.00– 1.30	Locked in cells
1.40– 4.00	Visits (if any)
2.00– 3.00	Exercise
4.00– 4.45	Tea
8.30	Cells locked for night

and at Parkhurst:

8.10	Cells unlocked, breakfast
10.00–11.30	Visits (if any)
10.30–11.30	(but now subject to variation) Exercise
12.00– 1.10	Locked in cells, lunch
1.30– 4.30	(but now subject to variation) Exercise
2.00– 4.15	Visits (if any)
5.00– 5.45	Locked in cells, tea
6.00– 7.00	(summer only and now subject to variation) Exercise
9.00	Cells locked for night

Classes take place during association and exercise periods. Exercise is optional and prisoners who choose not to exercise are free instead to use the recreational facilities within the unit.

Contact with families and friends outside is at least as important to most prisoners as any of the activities that occur while they are in prison. Indeed some prisoners will spend an enormous amount of their time writing letters or looking forward to visits. Anyone who knows anything about prisons knows how much visits and letters are valued and how much emotional energy may be invested in the quality of visits and the response to letters sent to loved ones. Prisoners in the security units, who are so closely confined and with such limited opportunities for personal contact, seem, unsurprisingly, to value such communication even more than men not held in such conditions.

At both prisons visits take place in visiting suites physically separate from the units and dedicated exclusively to that use, and at both prisons the inmates themselves arrange a visits rota. This is arrived at by agreement and apparently without difficulties. All visitors must be officially approved. Visits are conducted according to special arrangements which apply to all high escape risk Category A inmates but more visits are allowed than other prisoners receive. At Leicester there is a maximum of four two hour visits a month while at Parkhurst the most frequently visited prisoner has visitors twice a month. There is no set maximum but the location of the prison reduces demand. The Irish prisoners, in accordance with prison rules for inmates whose visitors cannot travel regularly, are among those who have what are known as 'accumulated visits', being visited perhaps each day of one week and then not for several months. Staff are required to maintain special vigilance during visits since there is always the risk of the importation of items that would compromise security. Thus they must observe the visit closely, and follow a number of other rules laid down for the secure conduct of visits. At the same time, staff and prisoners made clear to me that much effort is taken by staff to behave as sensibly as possible; for example visitors sometimes arrive with items forbidden by the rules but refusals are often effected informally in order to minimise fuss. Again, it is generally recognised that visits must be as rewarding as possible, that visiting arrangements must include the certainty that the visit will be accorded great respect by staff, and that any misbehaviour by one man on visits will not be followed by generalised punishment for all. Visits matter so much that they have to be treated as almost sacrosanct.

Letters too seem to assume even greater importance for security unit inmates than for most other prisoners. Each prisoner is issued one written letter per week at public expense; thereafter, at Leicester, letters must be paid for from earnings and prisoners complained that this was unreasonable and pointed out that they had not always been expected to pay for such extra letters. At Parkhurst, prisoners are allowed about five letters per week at public expense; if they exceed this they pay 5p towards each letter. All postage is paid at second class rate. Unit staff are informed by the prison security officer if the mail censors read of bad news likely to upset an inmate. This enables staff to prepare themselves for the prisoner's reaction and his subsequent behaviour. Staff mentioned that an inmate who is worried about something in the unit—such as his relationship with other inmates—will sometimes refer to it in a letter; he will know that his letter will be read by the mail censor, and it will be his way of informing staff of his concern without making an open approach. As already mentioned, some prisoners write extensively, and one current inmate has 143 correspondents.

There are of course other contacts with the outside than visits and letters. Radio, television and newspapers are readily available and are generally in much demand. In the United States many prisoners are allowed access to telephone and pay phones have now been introduced in open (Category D) prisons in England and Wales. This facility is being extended to Category C prisoners. Prisoners in higher security categories are sometimes allowed a telephone call if there are strong compassionate grounds, but there is no prospect of security unit inmates having routine access to telephones.

To sum up, the special security units offer a reasonable range of activities and are fairly generous in terms of visiting and letter-writing arrangements. In the latter respect they compare favourably with dispersal prisons. However, when account is taken of the more limited social opportunities and space, the overall regime is certainly no more attractive than imprisonment elsewhere. The purpose of the units, it is again worth stating, must be security first and then control. Security is achieved by demonstrating to prisoners that there is no point in trying to escape, control by providing a range of activities that occupy their time constructively and, as we shall see in the next chapter, by the quality of staff-inmate relationships. Unnecessary harshness would contribute nothing to security and be counter-productive in respect of control.

8 Relationships and control

Reference has been made in the last two chapters to the importance of regime structure, activities and letters and visits in connection with the maintenance of control. Prisoners confined in small security units with high staffing levels and opportunities for contacts with only a very small range of people require a comparatively unrestricted framework, a reasonable range of activities and relatively generous visiting and letter-writing arrangements if their frustration is not to boil over into disruption. These were some of the lessons learned in the early days of the special security units—especially from the history of the Durham unit. But perhaps more important for control than structures and activities, and at least as important as good arrangements for letters and visits, are the relationships in the unit, both among the inmates and between inmates and staff.

A key aspect of inmate-inmate relationships is considered to be the balance of different types of prisoner, frequently referred to as 'the mix'. Governors and prisoners alike stressed that loners, homosexuals and child murderers generally produced tension in the unit and that too many young men together created a volatile situation with the risk of 'macho' manifestations leading to trouble. And although half of the current inmates have been associated with Irish republican bombing campaigns, prison service personnel considered it undesirable that these men should outnumber other prisoners in either unit. As one former unit governor put it several years ago "experience has shown that it is absolutely crucial to have a good spread of inmates who are capable of providing a balance of power. At present we have three Irishmen, two younger prisoners and two older ones. This just about maintains a satisfactory mix". There is general uncertainty as to whether the harmony which generally prevails nowadays in the security units would withstand a significant alteration in the balance of unit populations, such as an influx of more younger people, or of a group of terrorists. What is agreed is that as far as possible the balance must be preserved. But with such considerations playing no part in the selection of individuals for security unit conditions, and with only two units currently available, allocation options are limited (see chapter 5 page 28 and 'the mix' is inevitably somewhat fortuitous.

It is also considered that the size of the unit populations is a relevant factor in control. The Parkhurst unit was built to hold a maximum of 20 inmates and coped with 17 on several occasions between March 1967 and February

1968, and the Leicester unit was built to hold 11 and held 9 at the end of 1966 and at various times between then and early 1973. But it was around that time that the units were at their most unsettled and the maximum capacity since 1980 has been officially regarded as 17 (10 at Parkhurst and 7 at Leicester). The governors of the two prisons feel that any more than these maxima would put unacceptable strains on the regimes in their units. It is also said that the smallness of the inmate groups in each unit, like the numerical superiority afforded by the manning levels, acts as a deterrent to disruptive behaviour.

Inevitably there are from time to time mutual antagonisms within the units. It is not only loners, homosexuals and child murderers who can be unpopular. There has at time been friction between different Irish republican groups, between representatives of rival London gangs and on one occasion between unit members and the colleague they no longer wanted to be cook. Living closely together, almost like a family, this is hardly surprising, and if antagonisms exceed a certain level or become entrenched a prisoner can be transferred to the other unit. The problems of living closely together are of course exacerbated for a man isolated from the group, unless he himself prefers to be alone. Equally it is difficult to cope with feelings of anger or frustration and occasional outbursts inevitably occur. Men can retreat to their cells to recover and will be allowed to do so, by prisoners and staff alike. Sometimes the cause of the flare-up will be a misunderstanding or a detail which has come to assume exaggerated importance in the closely confined conditions. As a group they have to have the ability to make up after a quarrel, as they would in a family relationship, and they must be able to apologise without looking weak. All antagonisms and all frustrations have to be subdued into a philosophy of live and let live. The cohesion of the prisoner group is highly valued for, as Sykes (1958) said in referring to inmate solidarity, it counteracts the pains of imprisonment.

The importance of power relationships among prison inmates is well known. It is noticeable in the units at present that the Irish republican bombers tend to hold dominant roles among the prison groups. This is partly an effect of their numbers, partly through force of personality, partly perhaps due to their ardent commitment to a cause, partly because such men are used to being in charge, and perhaps also partly because of an underlying awe of the power of the organisation to which they belong. But the units have always had leaders long before the arrival of the first bombers. And governors and unit staff do not believe that dominance in such a small family-type setting can long be sustained by bullying and fear. The dominant person is simply the leader who emerges, and he will often become spokesman for the prisoners and contribute to their cohesion by being able to help in the resolution of any differences between his fellows. But while the leader may not lead through being a bully or instilling fear, there is undoubtedly strong community pressure on occasions to conform in certain ways, even ways that may result in the loss of privileges, for example by participating in illegal activities in celebration of the anniversary of the death of Irish republican hunger-striker Bobby Sands.

As the Control Review Committee put it (1984) "relationships between staff and prisoners are at the heart of the whole prison system" and "control and security flow from getting that relationship right". Despite the difficulties of the 60s, periodic problems in the 70s and considerable tension and disruption after the reopening of the Leicester unit in 1980, staff-inmate relationships have now been consistently good for several years. This takes some explaining in that staff are the visible and ever-present representatives of the authority that has confined the unit inmates for many years in highly unpleasant circumstances. Prisoners cannot fail to be aware of what one described as "close-in control"—officers wherever you go, cameras and regular change-overs of staff.

The role of staff in the preservation of these good relationships must not be under-estimated. They have to perform a highly delicate task with great sensitivity. One inmate observed, undoubtedly with accuracy, that one officer can upset the atmosphere and that if relationships are not good there is a real potential for violence. At the Leicester unit the present position is that the bulk of staff chosen to work in the unit are mature, experienced officers; less experienced men may also be used but will work under close guidance from their superiors. It is regarded as important that those working in the unit want to work there. At Parkhurst all staff undertake duties in the unit in strict rotation by agreement with the local branch of the Prison Officers Association. At Leicester all unit staff, including the senior and principal officers do a stint of six months; continuity is maintained by principal and senior officers having designated reliefs who take over from them at the end of the six month period when new reliefs are designated. The changeover of principal officers and senior officers is at different times. The changeover of the officers is also staggered and they too are introduced to the unit by undertaking relief duties before their six month stint commences. At Parkhurst, principal and senior officers normally do 12 months in the unit and basic grade officers six months. The changeover of staff here too is a continuous one, with a small number of staff changed every month. These arrangements at both Leicester and Parkhurst are the result of experience; in earlier days longer periods of duty were undertaken and changeovers unstaggered.

Under current arrangements there is no standardised pattern either of staff training for work in the units or of staff support during their stints. With regard to training, reference has been made above to how officers are chosen and to the close supervision afforded to the less experienced staff at Leicester. It would probably be fair to say that working in the units requires an adaptation of normal working style rather than a complete change of methods. As for staff support, at times of continuous tension in the units this has been extremely important, but it is thought less essential today. At Leicester there used to be a support group meeting held once a week and chaired by the governor. This gave staff an opportunity to present any particular problem emerging in the unit and the governor used the meeting to make staff aware of his policy relating to the management of the unit. Currently there is no group support in the form of a regular meeting with a senior member of staff but support is available from a prison psychologist, and the governor visits the unit every day thus providing an opportunity for discussion of any matters of special concern. At Parkhurst there used to be a unit meeting every month chaired by the deputy governor. It was felt at the time that this gave staff an opportunity of sorting out problems and standards, increased staff group identification and enabled staff to learn both of the deputy governor's perception of developments in the unit and also of events occurring elsewhere in the prison of which they might otherwise have been unaware. Currently staff support is applied when it seems most needed. Budgeting restrictions are said to have played their part in the termination of the group, and in general the unit atmosphere is considered such as to render regular sessions unnecessary. Nonetheless, if tension were to persist in the unit or there were signs of a change in the atmosphere which might presage some trouble, I was told that steps would be taken to ensure that staff received adequate support. It is clearly important that staff working in the units are properly equipped to undertake the task and that the adequacy of training and support in this most sensitive of duties is regularly reviewed. Staff morale seems to be generally good despite the inevitably boring nature of what they are required to do.

At Leicester the four staff on duty at any one time in the unit are required to sit in twos at two different locations along the wing. These are their basic locations and from them they observe what is going on in the wing. Activity in cells, in the kitchen, the association room and other parts of the unit are monitored by periodic visits, and in practice the performance of these duties requires the staff to be fairly mobile. The principal officer will also move about freely within the unit. At Parkhurst a different method is used, with each of the activity areas being separately staffed. Thus there will be two officers in the gymnasium (unless it is not required and so locked) two in the association room, two in the workshop/arts and crafts room and so on. In both prisons the personal observations of unit staff are of course supplemented by the camera surveillance monitored by non-unit staff working in the electronic control room.

Such are the formal locational and observational duties of staff, the duties required by security considerations, but as already pointed out it is staff-inmate relationships which affect most vitally the atmosphere in the units and thus considerations of control. Governors and staff over the years have given much thought to the best way of conducting in a professional way their relationships with unit prisoners. There have been, and indeed still are, differing views as to how informal it is possible to be while still retaining control and, above all, remaining totally vigilant in matters affecting security. It is well recognised, for example in the Parkhurst unit's orders to staff, that because staff are in close proximity to a small number of prisoners over a long period it is inevitable that friendly relationships will develop with individuals. Most staff agree that it would be unreasonable and highly artificial to remain aloof and preserve a very distant relationship in such circumstances. Indeed prisoners are normally addressed by staff by their first names although it is unusual for staff to be so addressed by prisoners. (Staff are sometimes reluctant to admit outside the unit that first names are used, fearing that colleagues in the main prison would not understand.) It becomes difficult at times for staff to remember to have professional relationships. On the other hand some staff do prefer to keep the men 'at arm's length' and some wing principals and governor grades do not encourage the use of first names. One experienced officer, not working in a unit, argued that "familiarity breeds contempt", that when a man is convicted he loses the right to be called 'Mister' and that control is severely compromised if first names are used. As he put it "you can't order 'John' back to his cell". This would certainly be a minority view among unit staff but there were clearly variations in how friendly it was considered desirable or wise to be and prisoners would undoubtedly become suspicious if staff tried to get too close and they would reject such a stance. Perhaps the best way of describing staff-inmate relationships is to say that the degree of familiarity varies over time and according to governors, wing principals and individual staff. It also varies according to the degree of tension in the unit and according to whether that tension is generated by prisoners' grievances, by inter-inmate relationship difficulties, or by individual personalities.

Staff provide inmates with a much needed and usually much appreciated opportunity for conversation outside their small group of fellow-inmates. Prisoners frequently approach officers to chat and staff are usually encouraged to participate fully. Staff will also initiate conversations themselves. Being reasonably friendly, civilised and polite in one's dealings with prisoners, whatever their background and crime, is one of the hallmarks of the professionalism of the prison officer. It acknowledges for the prisoner his individuality and helps him to have confidence in his environment—for example that proper consideration will be given to his health and his welfare within the group. As one governor put it "he needs to know if there's a fire we will get him out". A prisoner is entitled not to be anxious for his personal safety and since any such anxiety is likely to make him more difficult to manage it is in the interests of staff to ensure that it is not present. Staff will sometimes extend their relationship with inmates into participating with them in games (eg snooker) and younger officers at the Parkhurst unit have been known to join in gymnasium activities.

The vital limitation on staff-inmate relationships is always security. It will be evident that there is a real danger that staff, used to easy-going friendly relationships with inmates and to participating in as civilised and rewarding a regime as possible, will relax their vigilance, especially if they are bored with their duties or if threats to security seem very remote. The Parkhurst unit's orders to staff encourage good relationships with inmates but go on to stress that "staff must *never* forget that their principal concern is security and should view all approaches by inmates with caution". Staff have always to guard against complacency and as one governor put it the danger with staff who are unfamiliar with dealing with notorious Category A prisoners is that "when they find they are not eaten alive by these fearful unknowns, they could switch off completely". It is felt that the two most serious escape attempts from the Leicester and Parkhurst units, in 1968 and 1976 respectively, owed much of their success to 'conditioning' of staff. In both units the regular turn-over of staff is regarded as an essential part of ensuring continued vigilance and unit managers know that they must be efficient at generating and maintaining good staff morale.

Despite the good staff-inmate relationships in the units, one factor, generally unspoken, is ever-present and governors and staff alike are well aware of it. The situation is redolent with potential threats to their careers and even their personal safety. There is always the risk that an officer will be compromised by accepting an inducement to show favour to a prisoner or prisoners which could by an on-going process of blackmail lead eventually to a security risk. The implications of falling victim to such a situation are of course understood by every member of staff. An apparently innocent request 'off the record' could eventually ruin their careers. Again staff must take care not to allow prisoners too much information about their personal lives; such information could be used to put pressure on staff which could likewise lead eventually to corruption. Security unit inmates are clever men and most of them have access to considerable financial and other resources. It is the 'other' resources which pose the underlying threat to governors and staff, since they know that trouble could be planned for them and their families. It is apparently very rare for threats to be received but knowledge of the possibility puts a pressure on them, especially when they feel bound to refuse a request in what they know to be a highly sensitive area, such as visits or matters bearing on political questions. Indeed the suggestion was put to me that this underlying threat may even contribute marginally to the positive attitude that staff feel able to adopt. Such a viewpoint seems unduly cynical and pays too little regard, I suspect, for the professionalism of staff. But the fact that the suggestion is even made underlines the extreme sensitivity of working with security unit prisoners.

Consideration of relationships within the units cannot properly ignore those between staff. Although staff are undertaking periods of only six months in the units, the small enclosed environment means that, just as prisoners have to learn to give and take in what is almost like a family situation, so officers too have to be flexible in order to ensure that they get on satisfactorily with each other. Flexibility of this kind is necessary in most work situations but it is especially important in working with a small group in unit conditions. If an officer were unable to fit in or unprepared to accept the regime he would be removed from the unit. It is in the interest of all that there should be no unnecessary rocking of the boat. Despite this, staff have some scope for arranging who will work together, subject to the overall requirements of the staff roster ('the detail').

Much stress has been laid on the importance of relationships within the units in connection with the maintenance of control. It has been noted that relationships are generally good and both units usually enjoy a relaxed atmosphere. It nonetheless is relevant to ask how often control problems do arise, what form they take and how they are handled.

Despite serious disruptions in the Durham unit in 1967 and 1968 and disturbances about the same time in the other units too, there has never been a repeat of trouble on this scale. The conditions, which were the prime sources of grievance, have been ameliorated, although they were implicated in the difficulties following the re-opening of the Leicester unit in 1980, and despite the militant opposition of prisoners in the Leicester unit to searching arrangements (1982) including a 'dirty protest', control problems have become rare. At Leicester the punishment cell has not been used for several years (Parkhurst does not have one) and at both units there are very few adjudications.

The comparative dearth of control problems does not surprise those who are familiar with the units. The prisoners in the units tend to be intelligent men who have chosen a particular way of life. Many verbalise well and can thus resolve their conflicts without recourse to violence. Unit prisoners tend to have a vested interest in retaining the current atmosphere and current facilities and not encouraging additional restrictions by making unnecessary difficulties. Indeed some keep an eye on others to ensure that trouble does not occur. Even in the units' early days this happened with men serving fixed prison sentences warning staff that if certain unstable lifers made trouble they would deal with them. By this manoeuvre they hoped to pre-empt a situation which could only have been to their own disadvantage: a man serving a fixed prison sentence can lose remission, while for a lifer with no prison release date such punishments may be less of a deterrent.

So there are few problems, and this is because unit prisoners are intelligent men with nothing to gain by causing trouble. But other factors are also mentioned as contributing to the predominantly trouble-free atmosphere. It is suggested that the men are resigned to their situation, that the manning levels deter trouble and that unstable people are rarely sent to the units. All these factors probably play their part, together with the determination of staff to avoid unnecessary problems and to handle the men as skilfully as possible. Reference has already been made to the sensitive refusal of forbidden presents from visitors. Some officers make a point of noting down everything they have undertaken to do for a prisoner, so that nothing should lead to their forgetting and thus upsetting men made more sensitive by their isolation.

Control problems, then, are now rare because prisoners have an interest in spending their days in a calm trouble-free environment and because the regime and the professionalism of staff are directed towards a similar objective. So long as the units are secure it is in everyone's interests that there should be no 'aggravation'.

But if there are fewer problems and this is unsurprising when all circumstances are taken into account, what is the nature and cause of such problems as do emerge? The most significant recent difficulties, according to both staff and prisoners, were occasioned by the presence in one unit of an inmate who in their view was inappropriately located in a security unit, being primarily a control problem rather than someone needing the very highest levels of security. This man had lost a considerable amount of remission during his time in the dispersal system, and unit staff and prisoners were equally proud that they had 'socialised' the man and that since his return to dispersal conditions he had now proved so 'reformed' that he had regained much of his lost remission. No other persistent control problem was quoted although tensions occasionally burst out into short-lived conflict. Almost anything is capable of being a source of conflict: inter-personal relationships, racial prejudice against Irish prisoners, religious or political disputes, catering, genuine misunderstandings, the inability to withdraw completely from company, visiting arrangements, insensitivities, bad news from home. Compulsory work was once a source of friction, as recorded earlier. In times past there has been violence against staff and against other prisoners, urine has been poured over staff and there was the 'dirty protest' mentioned above; and staff are always concerned about the possibility of hostage taking. Minor sources of grievance are not uncommon. In the Leicester unit, for example, prisoners would like an increase in their allowance from private cash to enable them to spend more on catering. But such matters do not normally disturb the generally untroubled atmosphere.

Such control problems as do arise are handled with all the sensitivity that staff can muster and on the basis of established relationships of mutual trust. Staff at both units take pride in their ability to cope with the prisoners held there and are able to accept occasional verbal abuse, followed frequently by an apology. At both units it is recognised that frustrations will sometimes boil over, and that punishing the resulting behaviour is rarely the appropriate response. If punishment is needed the prison rules allow loss of remission, cellular confinement, loss of association, loss of other privileges, loss of pay, and exercise taken separately from other prisoners. And, despite the rarity of such punishments, those who have had responsibility for the units are agreed on the need for suitable segregation facilities, outside the unit itself, at least to be available. At Leicester the main prison segregation unit is considered unsuitable for unit prisoners on security grounds and the punishment cell in the wing is regarded as inadequate for holding inmates for long periods. The only alternative is to confine a man to his cell. But this is not ideal since from his cell the inmate is still aware of all the wing activities and general noise and so cannot be afforded a period of quietness in which he can calm down completely separate from his normal environment. Also, if the other inmates in the wing were in sympathy with him, his presence in a locked cell could be a source of friction and disrupt the smooth running of the unit. One occasion is quoted in which an inmate had been confined to his cell, unjustifiably in the view of the others; they therefore decided not to have their usual video film to avoid upsetting him. At Parkhurst the situation is much the same. With no segregation facilities in the unit at all, not even a punishment cell, and the main prison's segregation unit, like Leicester's, regarded as unacceptable on security grounds, confinement in his own cell is the only option for serious misbehaviour. The drawbacks of taking such a step have already been outlined in the case of Leicester. The new security unit at Full Sutton prison will have a mini-segregation facility. As the governor of Full Sutton himself has written: "Whatever the degree of harmony arrived at by the majority, there always remains the problem of the 'outsider' who because of his personality, offences or political persuasions is rejected. Though experience suggests that resource to formal disciplinary procedures in . . . [the units] . . . is rare, so that the Full Sutton facility is unlikely to be used very much as a punishment unit, it will I am sure be valuable as a means of withdrawing from the community a man who has got himself into a state of conflict either with the prison authorities or with his fellow-prisoners. I anticipate that such withdrawals may occasionally take place on request, either from the man himself or from the others who need a rest from his behaviour".

This chapter has suggested that the atmosphere in the security units depends not only on the regime structure (buildings and management), on activities and on communications with the outside; relationships, both among the inmates and between inmates and staff are vitally important. In recent years there have been few control problems in the units and it is considered that 'the mix' of prisoners in each unit and the size of unit populations are both relevant to this. But a major factor is recognised as being the good staff-inmate relationships which have been developed as a result of sensitive staff handling of prisoners and also as a result of prisoners recognising that they too have an interest in serving their sentences in as tolerable an atmosphere as possible. At the same time, there is an underlying tension among staff that, as one governor put it, "if it suited them" (sc. the prisoners) "or things went wrong" (eg in an escape attempt) "good relationships would count for naught".

To conclude, and as an example of one aspect of staff-inmate relations, it is instructive to note the role played in one unit by the prisoners' leader and spokesman. He is in effect a buffer between the authorities and his fellow-inmates. Should a prisoner have a grievance or a request the leader will discuss it with him carefully and perceptively and the aggrieved will be pacified if there seems no genuine cause for complaint or he may be persuaded that pursuit of the grievance or request would inevitably be unproductive. On the other hand, if the leader accepts or indeed shares the grievance or is convinced that the request is justified, and also feels that that there is a real prospect of persuading the relevant authority (be it unit staff or governor) that the grievance/request could and should be resolved/met then he will plead the case. He will also weigh up what is to be gained or lost by pressing hard or not pressing at all for a particular point. It may be better to lose in one request and then use that outcome to bolster the case in respect of the next grievance/request. Such negotiation is seen by staff and the prisoners' spokesman himself as in the interests of both sides; and, quite clearly, his fellow prisoners also accept this analysis. Another example of the buffer role was when prisoners sensed that a key member of staff was unusually tense and uncommunicative. This in turn caused tension and anxiety among the prisoners. At length another member of staff was asked if the prisoners were perhaps seen as having done something wrong and it emerged that the explanation was simply that the member of staff in question was suddenly faced by a personal/domestic difficulty of some sort. The prisoners, on hearing this, relaxed and the 'family' could again cope, with prisoners as well as staff trying to be supportive to the member of staff concerned. The closeness of the small community could have seen the unusual reactions of one resulting in problems for the community as a whole, had this not been defused by clarification. Similarly and more commonly, the buffer role can help staff to understand why a particular prisoner is behaving unusually. Clearly the role is sensitive, necessitating the retention of trust by both sides. But it is a valuable one in which the prisoner concerned invests much and which he finds a constructive way of spending his time and energies.

9 The units and their host prisons

Having considered relationships within the units themselves it is important not to overlook the relationship between the units and their host prisons. Because the units are so 'special' within the prison system they inevitably differ greatly from the prisons within which they are situated. What effect does this difference have on the units and on the main prisons and how do governors and staff regard the presence of such a unit within the prison?

The Durham and Leicester units were established in 'local' prisons, prisons with a population mainly consisting of remand and short-sentence prisoners. However, Durham's E wing in which the unit was situated had recent experience of holding escape risks including long-sentence prisoners. The Chelmsford unit was established within a training prison containing short, medium and long-sentence prisoners. The Parkhurst unit was set up in a prison with a long record of holding prisoners serving long sentences and presenting serious security and control problems. With the creation of the dispersal system Parkhurst became one of the dispersal prisons. Leicester remains a 'local'. Full Sutton, where the new security unit will be, opened in October 1987 as a dispersal prison.

It might be assumed that the units' effects on their host prisons and also the effects of the host prisons on the units would vary according to the nature of the host prison, but there is no evidence that this is true to any significant extent. In fact, because of the separateness of the units within their host prisons and because of the priority that has to be given to them, there is little chance that the nature of the host prisons would greatly affect the units, although if staff in a unit based in a local prison were unable to adjust from working with the local prison population to working with long-term high security prisoners, or could not come to terms with the fact that as long-term prisoners the men in the security units receive more privileges than the local prison population, that would clearly have a negative effect. But there are no signs that the staff at the Leicester unit have such difficulties and it seems that both the Leicester and Parkhurst units proceed unaffected by the prison in which they are situated.

But what of the influence of the units on their host prisons? If priority must be accorded to a security unit, this would seem likely to have a negative effect on the rest of the prison. The danger of this occurring is recognised as being present. Indeed, if there was a major incident, such as an escape attempt, in the security unit, the governor might be faced with having to lock up the rest of the prison. But there have been few escape attempts and so a virtual absence of such disruption to the main prison.

Another potential source of difficulty for the main prison stems from the natural tendency for staff in a high profile unit to be seen as something of an elite and to provoke resentment and jealousies in other staff. But with the regular turnover of staff in both units and basic grade officers undertaking six month periods of duty there is little scope for such resentments to build up. One might have expected some friction at Leicester prison as a result of the fact that not all officers are chosen to work in the unit; that this does not seem to occur is probably because very few staff are so excluded.

There are nevertheless some matters associated with the units that periodically cause inconvenience or irritation to the main prisons at Leicester and Parkhurst. Staffing arrangements are one example. Because of the need to maintain in all circumstances the agreed staffing levels in the units any unexpected drain on staff resources must fall on the rest of the prison. Again, because the units require a number of staff and need them to be rotated at six month intervals it is not always possible to maintain the continuity of staffing that is desirable for the main prison. Apparently there have been extreme occasions when an officer doing an important job in the main prison is suddenly required to undertake his six months in the security wing; flexibility is clearly needed in such matters to avoid unnecessary difficulties.

Another source of irritation relates to the use of space and resources within the prison. At Leicester for example, the unit exercise yard, though small, occupies some 25% of the exercise space available to the whole prison; in a busy local prison this is not insignificant. Again, for security reasons vehicles inside Leicester prison are not allowed to pass between the prison wall and the yard while the prisoners are exercising. This is inconvenient.

Resentment has also been caused occasionally by the feeling that a disproportionate amount of resources/facilities was being allocated to the units. But it has not been a source of significant ill-feeling because of the recognition that unit prisoners have to be held in top security conditions.

There are other practical problems associated with the presence of a security unit. Extra media attention and interest from local politicians creates extra work. Management has to cope with the difficulties associated with preserving different regimes with different standards and objectives in the same prison; the changeover of staff could lead, for example, to some slackening in the stringency of the vigilance exercised by those manning the units. But this is not a major problem, simply a particular example of the task of management in any prison containing a variety of types of prisoner. Another such matter requiring management care is the slight danger of overlooking the interests of security unit personnel in discussions of the general management of the establishment; this is an effect of the separate and largely self-contained nature of the units. Again, at Parkhurst the use of a visiting-room outside the unit for all high-risk visits including those for men in the main prison necessitates co-ordination over the time-tabling; but this is little more than an administrative detail.

It is quite clear that all the negative aspects of having a security unit within a prison whose occupants are generally serving shorter sentences and require a lower level of security add up to very little. There is a potential danger of serious disruption to the prison but in practice the units are a source of no more than inconveniences one of which occasionally becomes a temporary source of significant irritation.

There seems to be a general consensus among staff and governors that any negative aspects to the presence of a security unit are insignificant compared to the positive features. Because of the special needs of the security units and the type of inmates contained there, staff working in the units feel that their position has acquired some status and they frequently mention their unit experience when going forward for promotion. It is said that almost without exception they respond very well to the high qualities of professionalism needed to work in such a place. It is not just a change for them from their normal routine in the main prison but something towards which most feel genuinely positive.

Governors too felt positive about the units and not just because of their good effect on staff. Just as working in a security unit is regarded by staff as a status job so the presence of such a unit in the prison has always been felt both at Leicester and Parkhurst to add a certain kudos to the whole establishment. The same was felt at Durham. The Leicester unit in particular is seen as having enhanced the prison in the city, no doubt because as a local prison it otherwise had a comparatively low profile. Governors do not regard its management as particularly taxing for them personally and it does not occupy a disproportionate amount of their time. There is generally a deputy governor with overall responsibility for the unit and a principal officer has charge of its day-to-day running; governors stress that most problems are solved at these levels and the skills of the PO have a direct bearing on the extent to which any problems are contained. Visiting the unit daily is part of the governor's normal routine and adjudications are few.

In one respect, the Leicester unit actually enhanced the regime of the local prison. When the unit opened, the prison was operating a very traditional regime without inmate association and with virtually no recreational activities other than a limited education programme and the occasional film show. The roofing-over of the end of one wing in order to house the unit provided a wide landing area well-suited to use for association. A system of limited association was accordingly introduced. So ironically, for structural reasons, the creation of the unit provided the basis for the introduction of association into Leicester prison for the first time.

Perhaps the overall conclusion should be that although there are some inconveniences and irritations associated with the presence in a prison of a special security unit these tend to be outweighed by positive features and the units and their host prisons really have very little effect on one another. Staff and governors are glad to have the units but generally, as one governor put it, "both bits go on oblivious of the other" and as another said "the general effect on [the host prison] was very little and I suspect that most staff were hardly aware of [the unit's] presence except when they were working in it".

10 Conclusions, lessons and the future

What then are the conclusions and lessons to be drawn from the foregoing account of the development and regimes of the special security units? This final chapter will be devoted to pulling together some of the threads. It will consider first some conclusions from the origins and development of the units; second, some lessons to be learned from the way the units are administered and managed, especially from the nature of the regimes; and third, some lessons from the experience of the security units that can be applied in the management of the new 'special units' that are being established for prisoners who stand out from the bulk of the prison population not by being the greatest *security* risks but by being among those who present the greatest *control* problems. Lastly, turning to the future, reference will be made to possible developments in respect of the units and their overall role in the prison system.

Origins and development

The special security units were born of the need to strengthen security for the most serious escape risks in the prison system, especially those with resources that might make them the object of a rescue attempt from outside. Three units containing 41 places were created between August 1965 and January 1966 at Durham, Leicester and Parkhurst prisons but by the time the third unit was open (Parkhurst) one of the others (Durham) had already been relegated to the second division in security and was to include prisoners who were the most violent and subversive in the system, in other words those who constituted the most serious control problems. Following the escape of the spy George Blake more men were drawn into the three units and a fourth emerged. Within 21 months of the first two units opening about 100 places were available in four units and 84 were filled. Here was a good example of the 'net widening' effect. The lesson is clear: if a new facility is created it may not be used exclusively for those for whom it was intended. Thus if it is regarded as important that it should be so used then there must be tight control on the process of selection. In the case of the Durham unit, it was administratively reasonable to take advantage of its existence to isolate the worst control risks from the rest of the system, there being no other convenient way of doing so. Today the security units are complemented by the 'special units' designed specifically for the constructive handling of prisoners who present the most serious control problems.

The conditions in these security units, however, were seen by Mountbatten (1966) as "Such as no country with a record of civilised behaviour ought to tolerate any longer than is absolutely necessary as a stopgap measure" and despite a number of improvements that were introduced the Home Office (1969) declared that "it remains undesirable that men should be detained for very long periods in such conditions". Further improvements have been made but conditions remain undesirable in that they are more restrictive than those provided for other prisoners detained over long periods. But the undesirable has been found to be necessary. Three men have spent over ten years in the security units and two of them were still there at the end of 1987.

In his report Lord Mountbatten recommended the categorisation of prisoners, with the greatest security risks labelled 'Category A'. He proposed that such men should be detained in a single fortress prison but the government subsequently decided to disperse them amongst a number of top security prisons and indicated that once the dispersal system was in being there would be no need for the special security units. However the first list of Category A prisoners contained four men not then held in the units for every three who were, and by the time six of the eight dispersal prisons were in operation at the end of 1970 the number of Category A prisoners in the units had only fallen from 61 to 40; it would thus have been reasonable to be sceptical as to whether the dispersal prisons would succeed in removing the perceived need for security units. But two units were indeed closed and by February 1972 the surviving units contained just 17 men. Within another couple of years the numbers might have been down to single figures had not the arrival on the scene of Irish republican bombers, and subsequently of other professional criminals with the resources that could lead to rescue attempts, removed the closure of further units from the immediate agenda. The changed policy —to retain the two remaining units—was formally agreed in 1978 and a decision was taken to build a new third unit which is to become operational in 1988.

The initial size and subsequent growth of the Category A population deserves special mention. Mountbatten argued in December 1966 that his fortress prison should be built for 120 men and that such an institution would eliminate any immediate need for a second prison. He clearly envisaged the initial Category A population being about 100 strong. In fact it started (mid 1967) at almost 150 and passed 200 in mid 1970. At the end of 1987 it stood at about 375. This raises the question whether from the start Category A was restricted to those Mountbatten had in mind, and whether in the last twenty years the criteria have become more lax. One dispersal governor said that he was convinced that Category A extends much wider than Mountbatten intended. It is also worth noting that compared with 375 Category A prisoners in England and Wales (population about 50 million, prison population about 50 thousand) Scotland (population about 5 million, prison population about 5 thousand) has just 10 Category A men. But Category A section at prison service headquarters are convinced that the criteria for entry into Category A have in fact become tighter and that many of the original Category A men of 1967 would not be so classified now. They argue that there are many more men now coming into the prison system whose escape would be 'highly dangerous to the public or police or to the security of the state' and draw attention to the presence among the 375 of some 50 members of Irish republican organisations, compared with none in Scotland and none in the original (1967) Category A list. The variation between the position in Scotland and that in England and Wales would seem to need closer examination but it is worth noting that the categorisation system in Scotland is not centralised, it being left to governors to make such decisions, and it is thought that Scotland would have many more Category A prisoners if the English categorisation system were used. Again, without a detailed comparison of Category A prisoners 20 years ago and Category A prisoners today, taking account of the type and severity of their offences and the length of their sentences, it is impossible to reach a conclusion as to whether the criteria have changed. Perhaps the main lesson to be drawn from the discrepancy between Mountbatten's expectations and what has happened since is that predictions relating to the prison population are extremely difficult to make and often incorrect.

Attention should also be drawn to one aspect of the way in which units have come to be closed. The wish to close a unit, first Durham, then Chelmsford, then (although it never happened) Parkhurst, in each case preceded decisions to reduce the population of the units. Instead of the process being

a. these men do not need to be in security units, let us disperse them;

b. that leaves too few in the units to justify retaining them all;

c. one unit can thus close.

it seems to have been rather

a. could a security unit be closed now?

b. on examination perhaps x men need not be in the units;

c. one unit can thus close.

This process of thought carries a risk that men may be retained in units longer than is necessary. Taken together with the evidence of net-widening in the early days of the units, it also raises the question whether the existence from 1988 of a third unit may not lead to a rise in the unit population merely because its very existence enables it to be used for the worst security risks not currently in the units. Clearly it is important that the increased provision should be justifiable on the present criteria of selection and should only be used to the extent that it is necessary; one of the three units could be converted into one of the new 'special units' for control problem prisoners if it becomes clear that there is a greater need for such facilities. There could also be a danger of the movement of prisoners in and out of the units being determined in accordance with administrative rather than security needs if the number of Irish republican bombers in the units (they comprised seven of the 14 inmates on 31.10.87) were to increase. Because of the desire to preserve a balance in each unit and avoid any unit being dominated by one group, there could be a temptation to retain other types of prisoner in the units longer than security demands, or to insert into the units additional prisoners who, had the 'mix' been otherwise, would not have been allocated to the units. Decisions would have to be taken as to whether to give in to such temptations on grounds of expediency—for operational reasons—or reject them on grounds of principle.

The prime conclusions to be drawn from an examination of the origins and development of the special security units seem to be these. They were necessary; the conditions though improved remain undesirable; they have successfully held their charges, losing only two men, one for two nights and one for two years; the rise in the number of prisoners with the resources to have a rescue organised on their behalf does not suggest that it will be possible for the prison system as presently constituted to manage without such units; nevertheless, care needs to be taken lest they expand unnecessarily.

Organisation and regime

Despite changes in the early days of the units, selection criteria have remained unchanged since February 1972 when the Leicester and Parkhurst units first stood on their own. There were then 17 inmates, there were 14 at the end of October 1987 and the population has never reached 20 throughout the intervening period. To meet the selection criteria a prisoner must be Category A and judged to be so great an escape risk that he cannot be contained safely within a dispersal prison. He will be so judged if it is thought that his escape would result in extreme danger (ie even greater danger than that posed by other Category A prisoners) to the public, the police or the security of the state or because he has come close to defeating the security of a dispersal prison or is believed to command the resources to do so.

The main changes to the unit population between 1972 and 1987 have been the arrival of the Irish republican bombers; the reduction in the number of other prisoners; the increased length of the sentences being served by inmates (with no current unit occupant serving less than 20 years); and a slightly increased length of stay in unit conditions because of the presence of the Irish bombers (from just over 5 years to 6½ years).

There is no formal regime policy for the units but there have been several attempts to establish a common regime. The buildings and space available vary between units and uniformity of regime is thus impossible. It has become clear that while inconsistencies should be eliminated as far as practicable—not least because of the sense of grievance this causes when inmates are transferred between the units—the really important consideration is that such differences as remain must be defensible. If the buildings and space in one unit are inferior it is recognised that it is reasonable for some compensating regime improvements to be introduced.

The regime framework may be seen as including the buildings and space, the staff manning levels, management practice and time unlocked. It is generally agreed that the physical conditions in both the Leicester and the Parkhurst unit leave much to be desired for long-term inmates, although the spacious exercise compound at Parkhurst is a valuable relief from the enclosed atmosphere, enabling prisoners to "get fresh air and sunshine and, perhaps more important, get away from each other". The new unit at Full Sutton will be the first unit to contain integral sanitation. The appropriate manning levels for a security unit have not been established to everyone's agreement but there is a case for experimenting with slightly lower levels where the nature of the buildings justify this, and such an experiment may be tried at Full Sutton, where a new purpose-built unit should create such an opportunity.

Management practice has developed over the years. It is now recognised that managers need to be strong, with a clear idea of what is wanted, but at the same time management must be unobtrusive with a highly professional approach and it must be well-controlled, creating no unnecessary aggravation for staff or prisoners. Everyone must know what is expected of them. It has become the accepted wisdom in the units, and it was very much the view of the Control Review Committee, that control is not effected by rigid disciplinary requirements but by the regime's activities, routines and procedures and above all by relationships, especially relationships between staff and inmates. A relaxed atmosphere and a degree of trust on both sides makes a positive contribution to control, and a flexibility with rules that do not bear on security and a willingness to give serious consideration to prisoners' requests invariably plays an important part in creating and maintaining good relationships and a good atmosphere.

Governors, staff and administrators have reached the above conclusions about the need for a relaxed regime and good relationships after having faced squarely the dilemma that this involves being reasonably friendly and positive towards men many of whom have committed appalling crimes. Such conclusions are based on recognition of the fact that the prime purpose of the units is security and the next most important objective is control. The units do not exist in order to make life unpleasant for the men although long term imprisonment in such confined conditions undoubtedly has that result, regardless of how the governor runs his prison. Keeping the regime as relaxed and humane as possible is in the interests of staff as well as prisoners. Unnecessary harshness would benefit neither. If the conditions obtaining in the early days of the units had not been improved, violence towards staff would inevitably have escalated and long-term solitary confinement would have been necessary, itself provoking widespread criticism as being inhumane. If the units can be made to work but experience shows that this depends on operating a regime that is more liberal than many would instinctively favour, the conclusion is that this is how they must be managed.

In line with the above considerations, prisoners in the special security units are allowed out of their cells for the maximum period possible. Because of the smaller numbers involved, this tends to be for about two hours longer than is allowed in the dispersal prisons. This may be seen as some compensation for the much more restricted environment in which they live. But the overall conditions are limited despite relaxed regimes and sensitive management by governors and staff.

Prison activities are now seen as significant in terms both of security and control. It is the control aspect which is most readily understood, since idleness clearly provides scope for trouble-making. But it is not hard to appreciate that boredom and frustration can equally encourage and provide additional opportunities for the planning of an escape. Despite this, compulsory work was found to be counter-productive in the security units, creating persistent control difficulties, including a general lack of co-operation, and being a source of much friction. Hobbies on the other hand, including some activities such as soft toy-making which were originally among the compulsory work options, are now recognised as an important means of enabling a prisoner to spend his time constructively. Food preparation has come to be seen as providing a good and much valued opportunity for self-expression and self-determination over a significant area of prison life. Education facilities are limited and because of the small numbers education has to be demand-led. Exercise is allowed for at least two hours a day and both units contain physical education equipment, though it is used by less than half the present inmates.

Contact with families and friends is at least as important to prisoners as any of the activities available. There is general agreement that it is in the interests of both staff and prisoners that every effort be made to ensure that visits are as rewarding as possible for inmates. This has to be achieved by sensitivity and good judgment since staff cannot afford to be generous in their interpretation of the rules for the secure conduct of visits. Letters too seem to assume even greater importance for security unit prisoners than for most others. It is not surprising that with security restrictions requiring the inmates' isolation not only from their loved ones but also from all but a small group of other prisoners and staff, such contacts with the outside as are possible will assume particular significance.

Thus, time has shown that prisoners confined in small special security units with high staffing levels and opportunities for contacts with only a very small range of people require a comparatively unrestricted framework, a reasonable range of activities and relatively generous visiting and letter-writing arrangements if their frustration is not to boil over into disruption.

But perhaps more important for control than structures and activities and at least as important as good arrangements for letters and visits are relationships in the units, both among inmates and between inmates and staff. Inmate-inmate relationships are agreed to depend partly on 'the mix' of prisoners with the requirement being "a good spread of inmates capable of providing a balance of power". But with such considerations playing no part in selection for special security unit conditions and with only two units currently available, allocation options are limited and the mix is inevitably somewhat fortuitous. Governors also believe that the smallness of inmate groups in each unit, like the numerical superiority afforded by manning levels, acts as a deterrent to disruptive behaviour.

As has been said, the Control Review Committee (1984) argued that "relations between staff and prisoners are at the heart of the whole prison system" and "control and security flow from getting that relationship right". Experience in the units has taught a number of lessons about getting that relationship right in the context of maximum security conditions. A relaxed atmosphere and mutual trust may be the foundation for success but there are one or two rather more practical points that have emerged as important. First, staff must do a limited tour of duty and handovers must be staggered. Overlong stints in such work create a risk that staff-inmate relationships may become too close and staff alertness to security needs weakened. Staggered handovers are necessary to ensure continuity. Second, attention must be paid to providing adequate training and support of staff working in this special environment. Third, staff, in forming good relationships with inmates must remain sufficiently detached to ensure that they are not compromised by accepting inducements to show favour and so leaving themselves open to blackmail; such care on the part of staff is of course necessary in any prison but the consequences of errors in a security unit are likely to be much greater, both in terms of security itself and the safety of the member of staff.

One factor which can aid control by contributing to good relationships between staff and inmates and indeed among inmates themselves is the role played by the dominant prisoner in a unit. If he is motivated towards reducing tensions and contributing to good relationships—and most unit prisoners do come to see that it is in their own interests that their long sentences should be served in as congenial an atmosphere as possible—then he can be an important influence for good.

In recent years there have been few control problems of note in the security units. Regime framework, activities, letters and visits arrangements, the mix, but above all atmosphere, relationships, trust, staff sensitivity; these appear to be the key to avoiding most difficulties and handling safely those that do occur. With few control problems, punishment is rarely employed but those who have responsibility for the units are agreed on the need for suitable segregation facilities to be available within the unit. Neither Leicester nor Parkhurst are so equipped although Full Sutton will be.

Present and past governors of the two existing units are agreed that the units work. They work in respect of their prime role of holding their inmates securely and they work in that prisoners and staff can co-exist in a reasonably good atmosphere. One governor summed up the reason the units work in the following way: "The men know the regime, the routine is clear, there is no ambiguity, staff and prisoners know each other, rules are consistent and faithfully adhered to, staff are firm but perfectly reasonable and reasonably friendly to all prisoners whatever their background and crime. Firm, fair, civilised, polite—most men will respond to this. You can thus avoid confrontation, conflict, tension. There is a certain identity of interest".

Lessons for the new 'special units'

Many of the lessons that have been learnt from the experience of the security units can equally be applied in the new 'special units' for long-term prisoners who present the severest control problems in the system. Others cannot. Security unit prisoners differ from severely disruptive prisoners in that they are in general more intelligent and less volatile; it is therefore likely that the security units will be more cohesive places with prisoners more able to calculate what is in their best interests and to act accordingly. But in both types of unit are men serving long sentences in relatively confined areas with relatively few fellow-prisoners for company. The following paragraphs draw attention to some of the lessons that do seem applicable in both types of unit. In addition, present and former governors at Leicester and Parkhurst, and also some of the current staff and inmates of the security units, gave advice on a number of aspects which they felt were important to the successful organisation and management of special units for control problem prisoners; their perceptive comments are quoted at various places.

The selection, allocation and transfer procedures for the special units differ from those used in the security units. The former are described at paras 121–5 of the report of the Research and Advisory Group (RAG) on the Long Term Prison System (1987), and selection is based on candidates referred by the governors of the prisons in which they are serving their sentence. Selection for the security units (described in chapter 5) is generally based on what is known about prisoners at the time of sentence. The special units are for men who show themselves to present serious control problems within the prison system; the security units are now (although chapter 1 showed it was not always so) mainly for men who are known to present serious security risks on the basis of what they have done before they entered the prison system. Only a minority of security unit inmates (4 out of the 14 inmates on 31.10.87) were not identified as requiring security unit conditions until some time into their sentence. These differences mean that security unit experience has little to teach those responsible for selection to the special units. Nor, since the first two special units are distinguished from each other by the fact that one is for men with a record of mental disorder while the other is not, are there yet any lessons to be learned in respect of allocation. But when the third special unit is opened at Hull towards the end of 1988, security unit experience suggests that allocators should have as much regard as possible to the 'mix' being created by allocations. It may be however that in the probably less cohesive groups of the special units, the question of mix is somewhat less important.

How long should men stay in the special units? Here again the security unit experience is of limited assistance. Security unit men must remain in such conditions so long as they are too great a security risk to be held in a dispersal prison. The purpose of placing a man in a special unit is where possible to change his behaviour for the better, but in many cases the realistic objective is to hold him in a positive environment while relieving the dispersal system of the disruption associated with his presence. The conclusion is perhaps that there should not be too great a pressure to get a man through and out of a special unit unless his behaviour has clearly improved; but at the same time, remembering the dangers of retaining security unit men longer than security requirements justify, care must be taken not to hold a special unit inmate longer than control requirements justify. One governor stressed also the need to establish criteria for release and to ensure that special unit inmates can clearly see that there is hope of advancing. The regular reviews of each special unit prisoner's progress, reviews in which he is himself involved, would seem to meet the latter requirement.*

It is felt within the prison service, both at headquarters and among governors well acquainted with the security units and the special units, that the establishment and regimes of the latter were more carefully planned than those of the security units. Certainly the pressure in 1964/5, following the rescue of one of the train robbers, was to make available without the slightest delay secure accommodation for men who might any day escape, and the urgency is well evidenced by the fact that another of the train robbers escaped just 5½ weeks before the accommodation was ready. The question what sort of regime to run was very much a secondary consideration.

Moves towards a common regime have been an important part of the history of the security units. Whether they will feature so prominently in the development of the special units remains to be seen. The special units are being planned to complement each other, though not to be an ad hoc collection of aims and regimes (see RAG 1987, paras 41–43). Diversity is thus part of the strategy. The lesson of the security units is surely that the planned differences should be known and understood by staff and should not be allowed to become inequitable. Further, every effort should be made to

*See also paras 54 and 55 of RAG 1987.

eliminate all differences which are not part of the strategy and all differences which are unnecessary.

The management of special units must surely be conducted on the same basis as that of security units. Staff must know exactly what the objectives are and what is expected of them. Security unit staff and prisoners also suggested that special unit prisoners should be clearly briefed. As one prisoner put it, they should be told: "this is what we are trying to do. You are in a last chance situation. You have been a disaster in the system but here it is up to you whether what we are trying works or fails". He also suggested that this involvement should be solicited by seeking their help perhaps in the running of some of the details of the regimes. The security unit experience has shown that the fostering of a relaxed atmosphere with no unnecessary aggravation creates the climate in which control problems are minimised. This means flexibility with rules where rigidity is not essential and a commitment to exercise control largely by the development of trust in staff and in the fairness of procedures. A willingness to give serious consideration to prisoners' requests and, for example, occasionally to grant an extra page for a letter above normal entitlements, is likely to be as conducive to increased trust and thus control in special units as it is generally felt to have proved itself to be in the security units.

The facilities available were felt by prison staff to be even more important for special unit prisoners than for those in the security units. It was pointed out that most security unit men have not been to a dispersal prison and thus are inclined to see the security unit regime as generous; by contrast all special unit men come from the dispersals and might be less willing to accept any limitations. Reference was made to the need for a routine that is respected and thus provides a framework on which prisoners can rely. There was general agreement that as in the security units as much time as possible should be spent out of cells. In fact, the security units currently manage a rather longer time unlocked—approximately 11 hours compared with 10¼ hours at Lincoln special unit and 8½ hours at the special unit at C wing Parkhurst. The difference between the 11 hours at the Parkhurst security unit and the 8½ hours at the special unit in the same prison is largely accounted for by staff 'communication sessions' and time needed for the writing of case-notes. Apparently, special unit inmates have shown no resentment about the time they spend locked up.

The regimes of special units should thus be characterised by similar features to those of the security units in terms of management and structure. Indeed a senior officer in one of the security units pointed out that in concentrating on the control aspects of a special unit it was important not to lose sight of the security side. And just as the management and structure of the two types of units should follow similar lines, so should the arrangements for activities and for letters and visits. The lessons that have been learnt about the priorities in all these matters in respect to security units were largely determined by considerations of control. They thus apply equally in special units where the reduction of control problems is the principal objective. Special unit prisoners, like their counterparts in the security units, are confined in small units with high staffing levels and opportunities for contacts with only a very limited range of people. Consequently they too require a comparatively unrestricted framework, a reasonable range of activities and relatively generous visiting and letter-writing arrangements if their frustration is not to boil over into disruption. After all, they have proved themselves seriously disruptive even in the less restricted environment of dispersal prisons.

The same considerations apply to the importance of good relationships between staff and inmates. Reference has been made on more than one occasion to the emphasis which the Control Review Committee (1984) put on this aspect as a means of reducing control problems. The Research and Advisory Group have made clear in their report on special units that they share that viewpoint (1987, paras 51–2, 83–4). Of the three particular lessons mentioned earlier as having been learnt in connection with ensuring appropriate staff-inmate relationships in the security units, two, in respect of the length of hours of duty in the units and the need to preserve a degree of detachment in relationships with inmates, were concerned with guarding against the possibility of security being endangered. Security is also very important in special units but the sharp differences between the two types of unit in the nature of their inmates must surely reduce the risks of it being endangered as a result of over-close relationships. The risks would seem no greater than in a dispersal prison, and at the C wing Parkhurst special unit it has not been found necessary to limit tours of duty in the way that is now regular practice in the security units. The third lesson identified was the importance of providing adequate training and support for staff working in a security unit. Governors and staff with security unit experience suggested that good staff support would be even more vital in a special unit, in view of the volatility of the inmates.

One major difference between the security units and the special units is that the former experience few control problems whereas the latter house only men who have presented serious control problems in more than one dispersal prison. In the handling of control problems as opposed to the avoidance of them there is only one lesson that the security unit experience teaches and that is in respect of physical structures: there is a need for suitable segregation facilities to be available whether these are within the unit or outside it.

Governors, staff and inmates of security units offered a number of additional comments about special units. In stressing the differences between security units and special units, they pointed out that the latter dealt with more difficult people but lacked the pressure that comes from handling men belonging to organisations that are powerful and dangerous. Despite the existence of personal officer schemes in the special units, which were seen by security unit inmates as a threat to the cohesion of the group (because of differential 'pushing power' of officers on behalf of their inmates), it was thought likely to be more difficult to form good relationships with special unit inmates; staff and inmates recalled the difficulty of getting on with one former security unit inmate who was seen by all as more of a control than a security problem. Again, several references were made to the difficulty of setting up special units in local prisons. Fears were expressed that in the early stages staff might, through inexperience of a group of difficult prisoners, revert to showing less understanding in their handling of the men than would be necessary. It was also feared that it would be more difficult in a special unit for staff and inmates to establish a relationship of mutual trust. And, to paraphrase the words of one prisoner, there are bound to be problems in living with each other and with officers; if relationships are not good there will be a real potential for violence, especially with prisoners who have a record of being difficult to control. Security unit prisoners approved unreservedly of the attempt to deal with special unit prisoners in the positive ways that have already been devised and that can perhaps be improved still further with experience (including the experience of security units); however, while wishing the special units well they were very sceptical of their chances of success.

What lessons for the special units can be drawn from the experience of the security units in respect of their relationships with their host prisons? As has been seen (chapter 9), despite some inconveniences and irritations associated with the presence of a security unit, these tend to be outweighed by positive features, such as the boost which working in a unit gives to staff and the kudos which the unit gives to the whole establishment. Overall the units seem to have little effect on the host prisons and the host prisons very little effect on the units. Is this likely to be the picture in the special units?

A crucial difference between the two types of unit stems from the superior status in prisons that security has over control. Security inevitably has the higher priority—keeping the prisoners in the prison is the prime duty of the prison service and is always going to be seen as more important than ensuring that they behave tolerably while they are there. Thus the status of units dealing with the top security risks is assured. And although control—maintaining good order in the interests of the operation of the prison and ensuring the safety of staff and inmates—is widely considered to be the second priority in prisons, it is natural that staff tend to have more mixed feelings about units for those who are seen as the top control problem prisoners. Prisoners who present the greatest control problems in the system do not have attached to them the 'glamour' of the top security risks, they can easily be seen as the most objectionable people in the system, and it is not surprising if some of the stigma such attitudes bring becomes associated also with those who work with them, especially if they are working with them in positive ways, based on creating a good atmosphere and good relationships. So one lesson may be that one cannot expect relationships with the host prisons to be as smooth in special units as in security units. Indeed, despite the evident early success of the special unit at Parkhurst C wing in relieving the dispersal system of a number of the most disruptive prisoners in the system and operating a positive regime without serious control difficulties, some tensions have emerged with the host prison, especially in respect of decisions about transfers of staff between the unit and the rest of the prison, some of which could be seen as showing insufficient commitment to the success of an important national resource (as Martin, 1987). The second special unit, which opened at Lincoln prison in May 1987, must face until it has been running for some time the difficulties of having to run a positive regime for difficult prisoners in a traditional local prison. As security unit staff foresaw, while staff are adjusting from working with the normal local prison population some problems are inevitable, although it must be said that initial signs at Lincoln are encouraging.

With both special units being staffed primarily by men with special training for the job and for a period substantially longer than the six months tours of duty in the security units there is also a much greater danger of resentments and jealousies between unit staff and their colleagues in the rest of the prison. When, as with the security units, almost all staff in the prison eventually do a tour of duty there is much less scope for undercurrents of feeling about unit staff being different and something of an elite. This danger clearly needs to be adequately addressed by management in the prisons in which the special units are located.

Another danger in special units is that unlike today's security units their very regime may be affected by the nature of the host prison. In particular a special unit situated in a local prison may find it difficult to see some of the most disruptive prisoners in the system accorded privileges akin to those in dispersal prisons when such privileges are considerably greater than those of local prisoners. However, just as in Leicester prison where the arrival of the security unit actually led to some improvements in the regime of the local prisoners (see chapter 9), so it may be that local prisons hosting special units may too gain some regime enhancement, perhaps as a result of a desire not to be outdone by the unit, perhaps as a result of the spread of some of the positive ideas behind the unit regime. Should there be a policy only to site special units in certain types of prison? One governor suggested that while it was perfectly sensible for security units to be located in dispersal prisons, such prisons were not a suitable location for special units. He argued that special units were intended to relieve the dispersal prisons of the aggravation caused by the most difficult men and that a special unit located within a dispersal failed to achieve this objective in that the men were still in the prison and the dispersal governor still had the responsibility for them. Current Home Office policy is not to select any particular type of prison for special units and until there is further evidence of their effects it is probably premature to conclude that one particular location is better than another.

The special units cannot therefore expect to have the status of the security units. However it would seem important that there should be a proper understanding of the difficulty of working closely with control problem prisoners and of the service to the dispersal system that special units are making in relieving it of many of those who disrupt things most and thus render the management of long-term prisoners—always one of the thorniest problems in any penal system—immeasurably more difficult. The role of the special units should be seen as being of very high significance for the successful running of the dispersal system and, arguably, almost on a par with the contribution made by the security units.

The future

The origins of the special security units are rooted in the realisation in the mid 1960s that there was then nowhere in the prison system sufficiently secure to withstand determined bids to rescue prisoners. The units continue in the late 1980s because, despite the development of the dispersal system, they alone are seen as capable of withstanding the most determined rescue attempts of today. Looking into the future, if the dispersal system, or some successor to that system—perhaps based on the 'new generation' designs that are now being introduced into some new prisons in this country—can be so arranged as to convince those to whom the responsibility falls that the units are superfluous then no doubt they will be closed. There are not likely to be many regrets since despite all the improvements to the conditions of the early days they leave much to be desired in units that hold men over long periods of time.

There are at present (December 1987) two units with a capacity of 17. There will shortly be three with a capacity of at least 25. How many places are needed and, if no alternative emerges, how many will be needed in the future? There is no easy answer to that question, except that as many places are and will be needed as there are men in prison whose escape would result in extreme danger to the public, the police or the security of the state, or who have come close to escaping from the most secure part of the rest of the prison system or are thought likely to have the resources to do so. Since 1972 that number has remained below 20. But if, as seems to be occurring, the number of prisoners who constitute an extreme danger and can command significant escape resources is increasing, and if the dispersal system is not rendered more secure (or replaced), then the need for security unit places is bound to grow and even 25 places may be insufficient. The important decision, then, if the number of such prisoners is indeed increasing, will be whether to strengthen (or replace) the dispersal system, to increase the number of security unit places, or to do both.

So long as the units remain they will need to be run in accordance with the lessons learned in their 22 year history, lessons particularly about the best management style, the use of activities and the importance of good staff–inmate relationships. There need not be a common regime but differences must be defensible. In this connection, one recent suggestion has been that the units might be jointly managed. Whether or not this is desirable, and the opinion of most governors and administrators seems to be that it is not, it would seem important that there be close co-operation between the governors in whose prisons the units are located and officials responsible for the administration of Category A prisoners, including a suitable forum for the joint discussion of issues of concern to them all.

Any penal system must make proper provision for those prisoners who constitute the gravest security risks. But it is right to judge a system not only by how securely it holds those men but also by how humanely it administers and manages the facility in which it holds them. There must also be the greatest care lest too many men be drawn into that facility and lest they be allowed to stay there too long. There is wide agreement that the special security units are necessary but, on grounds of economy and humanity alike, they should not contain more men than security demands and no prisoner should remain in a security unit for any longer than is absolutely necessary.

Appendix A Total occupancy of special security units 1965–87

	at 31 March	*at 30 June*	*at 30 September*	*at 31 December*
1965			15[1]	17[1]
1966	33[1]	43[1]	39	61
1967	53	82[2,3]	78[3]	77[3]
1968	71[3]	56[3,4]	48[5,6]	44[6]
1969	47[6]	45[6]	46	45
1970	41	40	39	39
1971	32	26	24[7]	25
1972	17[8]	16	14	13
1973	19	17	14	13
1974	14	13	13	15
1975	10	11	13	12
1976	13	11	11	6[9]
1977	6[9]	14	13	14
1978	9[10]	9[10]	8[10]	8[10]
1979	8[10]	9[10]	9[10]	9[10]
1980	10	11	13	13
1981	13	13	13	13
1982	13	13	12	13
1983	14	14	13	13
1984	12	12	12	14
1985	14	13	13	13
1986	13	15	12	12
1987	13	13	13	15

1. Includes one man held at Leicester unit by decision of local management and not a 'special security' prisoner.
2. Includes for the first time prisoners in Chelmsford unit.
3. Category A introduced June 1967—totals include non-Category A prisoners held in units.
4. Includes five convicted prisoners held in Brixton D wing and regarded as security unit prisoners. (See footnote at p. 16).
5. From September 1968 only Category A prisoners held in units.
6. Includes four convicted prisoners held in Brixton D wing.
7. Durham unit closed August 1971.
8. Chelmsford unit closed February 1972.
9. Parkhurst unit closed for security improvements.
10. Leicester unit closed for security improvements.

Appendix B Capacity and occupancy of individual units 1965–87 (summary)

The three original units were set up to hold 41 prisoners. There were 10 special security cells at Durham (opened 16.8.65), 11 at Leicester (also opened 16.8.65) and 20 at Parkhurst (opened 31.1.66).

Durham's 10 cells were all occupied by 12.11.65. Its capacity was increased to 40 by a confidential memorandum to governors of 20.5.66, but occupancy never exceeded 20 until after the escape of George Blake on 22.10.66. It reached 37 by 17.12.66 and its peak of 38 on 18.3.67. Following the disturbances at the prison in February 1967 and the immediate aftermath the numbers fell below 30 on 1.4.67 and remained at around 30 until the last week in July. They then fell to around 25 for the rest of the year and the early months of 1968. After the further disturbances in March 1968 the capacity was reduced, following a recommendation of the Inspector General, to a normal maximum of 13. At the beginning of February 1971 when the decision to close the unit was taken there were 14 prisoners (8 on the bottom landing and 6 on the top). The last prisoners left in August 1971.

Despite the provision of 11 cells at Leicester, no more than 9 were occupied at any one time. Papers of 1968 and 1969 indicated a maximum capacity of 10, although papers of 1971 and 1972 referred to 11 cells being available. Nine cells were occupied at various times between December 1966 and early 1973, but not since. Sometime between then and 1980 the capacity was reduced to 9 and it was reduced further to its current capacity of 7 late in 1980. An eighth cell is available for segregation/punishment.

Before Parkhurst's unit opened with 20 special security cells on 31 January 1966 a decision had been taken to occupy no more than 17 at any one time and to build up to that total gradually. Despite this the confidential memorandum of 20.5.66 indicated that 18 cells were available at Parkhurst. This had been reduced by 1.1.71 to 15 and by 1980 to 12. It was reduced to its current capacity of 10 late in 1980. The highest number held in the unit seems to have been 17 (March to June 1967, February 1968). It was last as high as 16 in October 1969, 15 in January 1970 and 13 in March 1970.

The fourth unit, at Chelmsford, had become officially recognised as a 'special security wing' by the beginning of May 1967. It could house at least 30 prisoners (as it did on 30.9.67) and there were 33 cells. However not all these cells were regarded as completely secure and it seems to have contained no more than ten Category A men until 1969. Normal capacity (for Cat A inmates) was regarded in May 1968 as eight, while by January 1969 it was regarded as ten with the expectation that it would rise to 20 when security improvements had been completed. By May 1969 capacity was 11 (expected to rise to 18) and when the closure decision was taken early in 1971 it was regarded as 20, although there were 23 cells in the unit.

The new unit at Full Sutton prison which is to be available in 1988 will have twelve cells of which two are for segregation.

Appendix C Key events in history of special security units

12.8.64	Escape of train robber Charles Wilson from Birmingham prison (recaptured 25 January 1968).
14.8.64	Home Secretary agrees to adapt parts of Durham, Leicester and Parkhurst prisons to provide special security conditions.
8.7.65	Escape of train robber Ronald Biggs from Wandsworth prison (not recaptured).
16.8.65	Units at Durham and Leicester open and receive prisoners.
20.11.65	Fears of armed rescue at Durham. Armed picquet stationed inside prison. (Withdrawn early February 1966 after transfer of train robbers.)
24.1.66	Governor of Durham informed that prisoners constituting the very highest degree of security risks would no longer be sent there.
31.1.66	Unit at Parkhurst opens and receives prisoners.
20.5.66	Memorandum to all governors announcing different criteria for Durham unit.
22.10.66	George Blake escapes from Wormwood Scrubs prison.
24.10.66	Lord Mountbatten appointed to conduct enquiry (reports 21 December).
8.2.67	Major disturbances at Durham unit ('the Football Mutiny').
1.5.67	Chelmsford unit sends in first official return of its occupants, as the other three units have done since May 1966.
8.6.67	Units reach peak occupancy level of 86.
22.6.67	Initial Category 'A' list issued (144 names, 61 of them in units).
3.3.68	Major disturbances at Durham unit ('the Chapel Barricade').
7.3.68	Radzinowicz committee recommends dispersal policy (dispersal programme announced in October 1968).

17.8.68	Major escape attempt from Leicester unit.
29.10.68	Three prisoners escape from Durham unit and one, John MacVicar from the prison (recaptured 11.11.70).
17.2.71	Announcement of forthcoming closure of Durham unit.
24 & 25.8.71	Last prisoners leave Durham unit (Ian Brady and John Straffen).
1.2.72	Announcement of forthcoming closure of Chelmsford unit.
16.2.72	Last prisoner leaves Chelmsford unit (Bruce Reynolds).
5.7.76	Three prisoners escape from Parkhurst unit and one, William Skingle, from the prison (recaptured 7.7.76).
26.11.76	Parkhurst unit closed for security improvements (reopened 12.4.77).
28.2.78	Leicester unit closed for security improvements (reopened 26.2.80).
1978	Ministerial decision no longer to phase out two remaining units and to seek funds from Treasury for a third unit (approval granted 1979).
1988	New unit to open at Full Sutton prison, Humberside.

Appendix D Work in special security units 1965–72

In November 1965, three months after the opening of the Durham and Leicester units prisoners at Durham were employed for 5½ hours a day sewing mailbags, with some rug-making undertaken also. There were two workshops, one on each of the two landings containing the special security cells. At Leicester work was for 2 hours in the morning and 2 hours in the afternoon, making lifebelts. It is recorded that six sewing machines were available for this work, which was undertaken each weekday and also on Saturday mornings.

The June 1967 instructions included four relating to work, as follows:

18. Economic viability will not be the principal consideration in providing work in the Special Security Wings, but work must be purposeful and interesting. Work requiring the use of tools may be provided.

19. The working day should be at least 6 hours. To achieve this in all wings it might be necessary for shift work to be arranged.

20. The cleaning of the Wing will be on a team work basis, but where it is necessary for cleaning to be continued during working hours, no prisoner should be employed as a cleaner for more than a week without the special directions of the Governor or the Medical Officer.

21. Wherever possible the decoration of the Wing should be carried out by the inmates themselves.

In November 1967 a progress report on work in the units showed that there were a number of developments in train and that the intention was to improve and expand the work opportunities. At Durham new workshops were being constructed with a view to introducing "new industries, that is wrought-iron and textiles". At the time however it was reported that inmates were employed on making lanyards or on painting toy soldiers. At Chelmsford work was to start shortly on the introduction of "the sign-making industry" but "fourteen men are at present engaged in making rugs" while "the other men are either engaged on the hand-sewing of mailbags or in doing the normal chores of the wing". It was noted that Chelmsford had "the only security wing workshop where the men are on an earnings scheme, and the Governor reports that they appear satisfied with the arrangements and are earning in the region of ten shillings a week each". At Leicester "the new workshop attached to the Special Security Wing'' was nearing completion and opening was due for January 1968. It would house ''a repair and maintenance operation for sewing machines''. The plan was then to depart from the procedure of splitting the inmates into two parties and to have only one party with a target working week of about 30 hours. The Governor considered it necessary to introduce an earnings scheme, as at Chelmsford, and sought a meeting with other Governors on this issue. At Parkhurst the new workshop was also due to be complete in January 1968. ''The main effort in the new workshop will be directed towards the repair and maintenance of sewing machines on the Island''. Again an earnings scheme was envisaged. It was also reported that approval had been given for the provision of a greenhouse—to be assembled by inmates during the winter. Category A women at Holloway, then regarded as a 'Special Security Wing' along with the other four, were working on daily chores, on the repair of linen and, in the case of two inmates on rule 43, on embroidery or tapestry work and ''making name brooches for the nursing sisters of the prison service''.

The improved arrangements for employment in the units were actually available in March 1968 and were as follows:

Chelmsford The manufacture of warning triangles—light engineering industry. Employing about 15 men. In addition rug-making continued on a hobby basis.

Durham Light textiles (employing 6 men) and light wrought-iron work (employing about 14). In addition, as part of hobby activities, soft toy making, toy painting.

Leicester Repair and maintenance of sewing machines (employing 6–8 men). Carpentry and joinery continued on a hobby basis.

Parkhurst Repair and maintenance of sewing machines (employing 9 men). Soft toy making and minor carpentry work continued as an alternative.

Reasonable work was thus available at all units in the Spring of 1968. But within a few months a suggested regime objective for the units was a clear forerunner of the ultimate decision to withdraw the requirement to work. The first notes prepared in July 1968 for discussion on a common unit regime saw three main objectives in respect of employment in the units, one of which was ''to clarify the inmates right to opt out of work if he doesn't feel like it''. The other two objectives were ''to standardise the length of the working day'' and ''to have a clear-cut policy on the relation of work to such other aspects of the programme as physical education, hobbies, counselling etc''. Three years later it was recorded (September 1971) that only ''a very small minority of prisoners 'work in the accepted sense'. They are nevertheless paid top rates for nominal cleaning duties''. This situation had arisen because workshop provision was still recognised as being more limited than that available for other long-term prisoners and because some unit offenders had to be kept separate from others (eg sex offenders on the upper floor at Durham). Compulsory work at the special security units was thus discontinued at about the time Leicester and Parkhurst became the only two surviving units (February 1972). In addition to the practical problems of providing suitable work, there were also the very important implications for control. The requirement to work gave rise to inmates refusing to do so, generated a general lack of co-operation and became a source of considerable friction.

References

Bolton, N. *et al.* (1976). Psychological correlates of long-term imprisonment. *British Journal of Criminology* 16, pp. 38–47.

Cohen, S. and Taylor, L. (1982). *Psychological survival.* London: Penguin.

Control Review Committee (1984). *Managing the long-term prison system* (report). London: HMSO.

Dunbar, I. (1986). *A sense of direction.* London: Home Office

Hansard (1965). Parliamentary Answer 25.11.65.

Hansard (1965). Home Office Questions 9.12.65.

Hansard (1965). Parliamentary Answer 16.12.65.

Hansard (1968). Parliamentary Answer 24.10.68.

Hansard (1971). House of Lords Announcement 17.2.71.

Hansard (1972). Parliamentary Answer 1.2.72.

Home Office (1962). *Report of the Commissioners of Prisons for the year 1961.* Cmnd 1798. London: HMSO.

Home Office (1967). *Report on the work of the Prison Department, 1966* Cmnd 3408. London: HMSO.

Home Office (1968). *Report on the work of the Prison Department, 1967* Cmnd 3774. London: HMSO.

Home Office (1969). *Report on the work of the Prison Department, 1968* Cmnd 4186. London: HMSO.

Home Office (1969). White paper '*People in Prison*'. Cmnd 4214. London: HMSO.

Home Office (1971). *Report on the work of the Prison Department 1970* Cmnd 4724. London: HMSO.

Home Office (1973). Working Party on '*Dispersal and Control*'. Unpublished report. Results announced to Parliament 11.5.73.

Home Office (1977). *Report on the work of the Prison Department 1976* Cmnd 6877. London: HMSO.

Home Office (1985). *Report on the work of the Prison Department 1984–5* Cmnd 9699. London: HMSO.

King, R. D. and Elliott, K. W. (1977). *Albany: birth of a prison—end of an era.* London: Routledge and Kegan Paul.

Martin, J. P. (1987). *C Wing HMP Parkhurst: some aspects of management.* Report to Home Office (as yet unpublished).

Morris, T. P. (1968). Review of Radzinowicz report. *British Journal of Criminology* Vol 8 No 3. London: Stevens.

Mountbatten (1966). *Report of the Inquiry into Prison Escapes and Security* Cmnd 3175. London: HMSO.

Radzinowicz, L. (1968). The regime for long-term prisoners in conditions of maximum security. *Report of the Advisory Council on the Penal System.* London: HMSO.

Research and Advisory Group (RAG) on the Long-Term Prison System (1987). *Special units for long-term prisoners.* London: HMSO.

Richards, B. (1978). 'The experience of long-term imprisonment.' *British Journal of Criminology,* 18, pp. 162–169.

Sapsford, R. J. (1978). 'Life sentence prisoners: psychological changes during sentence' *British Journal of Criminology,* 18, pp. 128–145.

Sapsford, R. J. (1983). *Life sentence prisoners.* Milton Keynes: Open University Press.

Sykes, G. M. (1958). *The society of captives.* Princeton N J: Princeton Univ.

Times (1970). 'The four security wings that are a blot on Britain's gaol system.' Article, 23 October 1970.

West, D. J. (1968). Records of 138 Category 'A' prisoners. Report summary at Appendix C of Radzinowicz L. (1968).

Publications

Titles already published for the Home Office

Studies in the Causes of Delinquency and the Treatment of Offenders (SCDTO)

1. Prediction Methods in relation to borstal training. Hermann Mannheim and Leslie T. Wilkins, 1955. viii+276pp. (11 340051 9).
2. *Time spent awaiting trial. Evelyn Gibson. 1960. v+45pp. (34-368-2).
3. *Delinquent generations. Leslie T. Wilkins. 1960. vi+20pp. (11 340053 5).
4. *Murder. Evelyn Gibson and S. Klein. 1961. iv+44pp. (11 340054 3).
5. Persistent criminals. A study of all offenders liable to preventive detention in 1956. W. H. Hammond and Edna Chayen. 1963. ix+237pp. (34-368-5).
6. *Some statistical and other numerical techniques for classifying individuals. P. McNaughton-Smith. 1965. v+33pp. (34-368-6).
7. Probation research: a preliminary report. Part I. General outline of research. Part II. Study of Middlesex probation area (SOMPA). Steven Folkard, Kate Lyon, Margaret M. Carver and Erica O'Leary. 1966. vi+58pp. (11 340374 7).
8. *Probation research: national study of probation. Trends and regional comparisons in probation (England and Wales). Hugh Barr and Erica O'Leary. 1966. vii+51pp. (34-368-8).
9. *Probation Research. A survey of group work in the probation service. Hugh Barr. 1966. vii+94pp. (34-368-9).
10. *Types of delinquency and home background. A validation study of Hewitt and Jenkins' hypothesis. Elizabeth Field. 1967. vi+21pp. (34-368-10).
11. *Studies of female offenders. No. 1—Girls of 16–20 years sentenced to borstal or detention centre training in 1963. No. 2—Women offenders in the Metropolitan Police District in March and April 1957. No. 3—A description of women in prison on January 1, 1965. Nancy Goodman and Jean Price. 1967. v+78pp. (34-368-11).
12. *The use of the Jesness Inventory on a sample of British probationers. Martin Davies. 1967. iv+20pp. (34-368-12).
13. *The Jesness Inventory: application to approved school boys. Joy Mott. 1969. iv+27pp. (11 340063 2).

Home Office Research Studies (HORS)

1. *Workloads in children's departments. Eleanor Grey. 1969. vi+75pp. (11 340101 9).
2. *Probationers in their social environment. A study of male probationers aged 17–20, together with an analysis of those reconvicted within twelve months. Martin Davies. 1969. vii+204pp. (11 340102 7).
3. *Murder 1957 to 1968. A Home Office Statistical Division report on murder in England and Wales. Evelyn Gibson and S. Klein (with annex by the Scottish Home and Health Department on murder in Scotland). 1969. vi+94pp. (11 340103 5).
4. *Firearms in crime. A Home Office Statistical Division report on indictable offences involving firearms in England and Wales. A. D. Weatherhead and B. M. Robinson. 1970. viii+39pp. (11 340104 3).
5. *Financial penalties and probation. Martin Davies. 1970. vii+39pp. (11 340105 1).

---

*Out of Print.

6. *Hostels for probationers. A study of the aims, working and variations in effectiveness of male probation hostels with special reference to the influence of the environment on delinquency. Ian Sinclair. 1971. ix + 200pp. (11 340106 X).
7. *Prediction methods in criminology—including a prediction study of young men on probation. Frances H. Simon. 1971. xi + 234pp. (11 340107 8).
8. *Study of the juvenile liaison scheme in West Ham 1961–65. Marilyn Taylor. 1971. vi + 46pp. (11 340108 6).
9. *Exploration in after-care. I—After-care units in London, Liverpool and Manchester. Martin Silberman (Royal London Prisoners' Aid Society) and Brenda Chapman. II—After-care hostels receiving a Home Office grant. Ian Sinclair and David Snow (HORU). III—St. Martin of Tours House, Ayreh Leissner (National Bureau for Co-operation in Child Care). 1971. xi + 140pp. (11 340109 4).
10. A survey of adoption in Great Britain. Eleanor Grey in collaboration with Roland M. Blunden. 1971. ix + 168pp. (11 340110 8).
11. *Thirteen-year-old approved school boys in 1962. Elizabeth Field, W. H. Hammond and J. Tizard, 1971. xi + 46pp. (11 340111 6).
12. Absconding from approved schools R. V. G. Clarke and D. N. Martin. 1971. vi + 146pp. (11 340112 4).
13. An experiment in personality assessment of young men remanded in custody. H. Sylvia Anthony. 1972. viii + 79pp. (11 340113 2).
14. *Girl offenders aged 17–20 years. I—Statistics relating to girl offenders aged 17–20 years from 1960 to 1970. II—Re-offending by girls released from borstals or detention centre training. III—The problems of girls released from borstal training during their period on after-care. Jean Davies and Nancy Goodman. 1972. v + 77pp. (11 340114 0).
15. *The Controlled trial in institutional research–paradigm or pitfall for penal evaluators? R. V. G. Clarke and D. B. Cornish. 1972. v + 33pp. (11 340115 9).
16. *A survey of fine enforcement. Paul Softley. 1973. v + 65pp. (11 340116 7).
17. *An index of social environment—designed for use in social work research. Martin Davies. 1973. vi + 63pp. (11 340117 5).
18. *Social enquiry reports and the probation service. Martin Davies and Andrea Knopf. 1973. v + 49pp. (11 340118 3).
19. *Depression, psychopathic personality and attempted suicide in a borstal sample. H. Sylvia Anthony, 1973. viii + 44pp. (11 340119 1).
20. *The use of bail and custody by London magistrates' courts before and after the Criminal Justice Act 1967. Frances Simon and Mollie Weatheritt. 1974. vi + 78pp. (11 340120 5).
21. Social work in the environment. A study of one aspect of probation practice. Martin Davies, with Margaret Rayfield, Alaster Calder and Tony Fowles. 1974. ix + 151pp. (11 340121 3).
22. Social work in prison. An experiment in the use of extended contact with offenders. Margaret Shaw. 1974. vii + 154pp. (11 340122 1).
23. Delinquency amongst opiate users. Joy Mott and Marilyn Taylor. 1974. vi + 31pp. (11 340663 0).
24. IMPACT. Intensive matched probation and after-care treatment. Vol. I—The Design of the probation experiment and an interim evaluation. M. S. Folkard, A. J. Fowles, B. C. McWilliams, W. McWilliams, D. D. Smith, D. E. Smith and G. R. Walmsley. 1974. v + 54pp. (11 340664 9).
25. The approved school experience. An account of boys' experiences of training under differing regimes of approved schools, with an attempt to evaluate the effectiveness of that training. Anne B. Dunlop. 1974. vii + 124pp. (11 340665 7).
26. *Absconding from open prisons. Charlotte Banks, Patricia Mayhew and R. J. Sapsford. 1975. vii + 89pp. (11 340665 5).
27. Driving while disqualified. Sue Kriefman. 1975. vi + 136pp. (11 340667 3).
28. Some male offenders' problems. I—Homeless offenders in Liverpool. W. McWilliams. II—Casework with short-term prisoners. Julie Holborn. 1975. x + 147pp. (11 340668 1).
29. *Community service orders. K. Pease, P. Durkin, I. Earnshaw, D. Payne and J. Thorpe. 1975. viii + 80pp. (11 340669 X).
30. Field Wing Bail Hostel: the first nine months. Frances Simon and Sheena Wilson. 1975. viii + 55pp. (11 340670 3).

---

*Out of Print.

31. Homicide in England and Wales 1967–1971. Evelyn Gibson. 1975. iv+59pp. (11 340753 X).
32. Residential treatment and its effects on delinquency. D. B. Cornish and R. V. G. Clarke. 1975. vi+74pp. (11 340672 X).
33. Further studies of female offenders. Part A: Borstal girls eight years after release. Nancy Goodman, Elizabeth Maloney and Jean Davies. Part B: The sentencing of women at the London Higher Courts. Nancy Goodman, Paul Durkin and Janet Halton. Part C: Girls appearing before a juvenile court. Jean Davies. 1976. vi+114pp. (11 340673 8).
34. *Crime as opportunity. P. Mayhew, R. V. G. Clarke, A. Sturman and J. M. Hough. 1976 vii+36pp. (11 340674 6).
35. The effectiveness of sentencing: a review of the literature. S. R. Brody. 1976. v+89pp. (11 340675 4).
36. IMPACT. Intensive matched probation and after-care treatment. Vol II—The results of the experiment. M. S. Folkard, D. E. Smith and D. D. Smith. 1976. xi+400pp. (11 340676 2).
37. Police cautioning in England and Wales. J. A. Ditchfield. 1976. vi+31pp. (11 340677 2).
38. Parole in England and Wales. C. P. Nuttall, with E. E. Barnard, A. J. Fowles, A. Frost, W. H. Hammond, P. Mayhew, K. Pease, R. Tarling and M. J. Weatheritt. 1977. vi+90pp. (11 340678 9).
39. Community service assessed in 1976. K. Pease, S. Billingham and I. Earnshaw. 1977. vi+29pp. (11 340679 7).
40. Screen violence and film censorship: a review of research. Stephen Brody. 1977. vii+179pp. (11 340680 0).
41. Absconding from borstals. Gloria K. Laycock. 1977. v+82pp. (11 340681 9).
42. Gambling: a review of the literature and its implications for policy and research. D. B. Cornish. 1987. xii+284pp. (11 340682 7).
43. Compensation orders in magistrates' courts. Paul Softley. 1978. v+41pp. (11 340683 5).
44. Research in criminal justice. John Croft. 1978. iv+16pp. (11 340684 3).
45. Prison welfare: an account of an experiment at Liverpool. A. J. Fowles. 1978. v+34pp. (11 340685 1).
46. Fines in magistrates' courts. Paul Softley. 1978. v+42pp. (11 340686 X).
47. Tackling vandalism. R. V. G. Clarke (editor), F. J. Gladstone, A. Sturman and Sheena Wilson (contributors). 1978. vi+91pp. (11 340687 8).
48. Social inquiry reports: a survey. Jennifer Thorpe. 1979. vi+55pp. (11 340688 6).
49. Crime in public view. P. Mayhew, R. V. G. Clarke, J. N. Burrows, J. M. Hough and S. W. C. Winchester. 1979. v+36pp. (11 340689 4).
50. *Crime and the community. John Croft. 1979. v+16pp. (11 340690 8).
51. Life-sentence prisoners. David Smith (editor), Christopher Brown, Joan Worth, Roger Sapsford and Charlotte Banks (contributors). 1979. iv+51pp. (11 340691 6).
52. Hostels for offenders. Jane E. Andrews, with an appendix by Bill Sheppard. 1979. v+30pp. (11 340692 4).
53. Previous convictions, sentence and reconviction: a statistical study of a sample of 5,000 offenders convicted January 1971. G. J. O. Phillpotts and L. B. Lancucki. 1979. v+55pp. (11 340693 2).
54. Sexual offences, consent and sentencing. Roy Walmsley and Karen White. 1979. vi+77pp. (11 340694 0).
55. Crime prevention and the police. John Burrows, Paul Ekblom and Kevin Heal. 1979. v+37pp. (11 340695 9).
56. Sentencing practice in magistrates' courts. Roger Tarling, with the assistance of Mollie Weatheritt. 1979. vii+54pp. (11 340696).
57. Crime and comparative research. John Croft. 1979. iv+16pp. (11 340697 5).
58. Race, crime and arrests. Philip Stevens and Carole F. Willis. 1979. v+69pp. (11 340698 3).
59. Research and criminal policy. John Croft. 1980. iv+14pp. (11 340699 1).
60. Junior attendance centres. Anne B. Dunlop. 1980. v+49pp. (11 340700 9).
61. Police interrogation: an observational study in four police stations. Paul Softley, with the assistance of David Brown, Bob Forde, George Mair and David Moxon. 1980. vii+67pp. (11 340701 7).
62. Co-ordinating crime prevention efforts. F. J. Gladstone. 1980. v+74pp. (11 340702 5).

---

*Out of Print.

63. Crime prevention publicity: an assessment. D. Riley and P. Mayhew. 1980. v+47pp. (11 340703 3).
64. Taking offenders out of circulation. Stephen Brody and Roger Tarling. 1980. v+46pp. (11 340704 1).
65. *Alcoholism and social policy: are we on the right lines? Mary Tuck. 1980. v+30pp. (11 340705 X).
66. Persistent petty offenders. Suzan Fairhead. 1981. vi+78pp. (11 340706 8).
67. Crime control and the police. Pauline Morris and Kevin Heal. 1981. v+71pp. (11 340707 6).
68. Ethnic minorities in Britain: a study of trends in their positions since 1961. Simon Field, George Mair, Tom Rees and Philip Stevens. 1981. v+48pp. (11 340708 4).
69. Managing criminological research. John Croft. 1981. iv+17pp. (11 340709 2).
70. Ethnic minorities, crime and policing: a survey of the experiences of West Indians and whites. Mary Tuck and Peter Southgate. 1981. iv+54pp. (11 340765 3).
71. Contested trials in magistrates' courts. Julie Vennard. 1982. v+32pp. (11 340766 1).
72. Public disorder: a review of research and a study in one inner city area. Simon Field and Peter Southgate. 1982. v+77pp. (11 340767 X).
73. Clearing up crime. John Burrows and Roger Tarling. 1982. vii+31pp. (11 340768 8).
74. Residential burglary: the limits of prevention. Stuart Winchester and Hilary Jackson. 1982. v+47pp. (11 340769 6).
75. Concerning crime. John Croft. 1982. iv+16pp. (11 340770 X).
76. The British Crime Survey: First Report, Mike Hough and Pat Mayhew. 1983. v+62pp. (11 340789 6).
77. Contacts between police and public: findings from the British Crime Survey. Peter Southgate and Paul Ekblom. 1984. v+42pp. (11 340771 8).
78. Fear of crime in England and Wales. Michael Maxfield. 1984. v+51pp. (11 340772 6).
79. Crime and police effectiveness. Ronald V. Clarke and Mike Hough. 1984. iv+33pp. (11 340773 4).
80. The attitudes of ethnic minorities. Simon Field. 1984. v+50pp. (11 340077 2).
81. Victims of crime: the dimensions of risk. Michael Gottfredson. 1984. v+54pp. (11 340775 0).
82. The tape recording of police interviews with suspects: an interim report. Carole Willis. 1984. v+45pp. (11 340776 9).
83. Parental supervision and juvenile delinquency. David Riley and Margaret Shaw. 1985. v+90pp. (11 340799 8).
84. Adult prisons and prisoners in England and Wales 1970–82: a review of the findings of social research. Joy Mott. 1985. vi+73pp. (11 340801 3).
85. Taking account of crime: key findings from the 1984 British Crime Survey. Mike Hough and Pat Mayhew. 1985. vi+115pp. (11 340810 2).
86. Implementing crime prevention measures. Tim Hope. 1985. vi+82pp. (11 340812 9).
87. Resettling refugees: the lessons of research. Simon Field. 1985. vi+62pp. (11 340815 3).
88. Investigating burglary: the measurement of police performance. John Burrows. 1986. v+36pp. (11 340824 2).
89. Personal violence. Roy Walmsley. 1986. vi+87pp. (11 340827 7).
90. Police public encounters. Peter Southgate with the assistance of Paul Ekblom. 1986. vi+150pp. (11 340834 X).
91. Grievance procedures in prisons. John Ditchfield and Claire Austin. 1986. vi+78pp. (11 340839 0).
92. The effectiveness of the Forensic Science Service. Malcolm Ramsay. 1987. v+100pp. (11 340842 0).
93. The police complaints procedure: a survey of complainants' views. David Brown. 1987. v+98pp. (11 340853 6).
94. The validity of the reconviction prediction score, Denis Ward. 1987. vi+40pp. (11 340682 X).
95. Economic aspects of the illicit drug market and drug enforcement policies in the United Kingdom. Adam Wagstaff and Alan Maynard. 1988. vii+156pp. (11 340883 8).
96. Schools, disruptive behaviour and delinquency: a review of research. John Graham. 1988. v+70pp. (11 340887 0).

---

*Out of Print.

97. The tape-recording of policy interviews with suspects: a second interim report. Carole Willis, John Macleod and Peter Naish. 1988. vii + 97pp. (11 340888 9).
98. Triable-either-way cases: Crown Court or magistrates' court? David Riley and Julie Vennard. 1988. v + 52pp. (11 340890 0).
99. Directing patrol work: a study of uniformed policing. John Burrows and Helen Lewis. 1988. v + 66pp. (11 340891 9).
100. Probation day centres, George Mair. 1988. v + 44pp. (11 340894 3).
101. Amusement machines: dependency and delinquency. John Graham. 1988. v + 48pp. (11 340895 1).
102. The use and enforcement of compensation orders in magistrates' courts. Tim Newburn. 1988. v + 48pp. (11 340896 X).
103. Sentencing practice in the Crown Court. David Moxon. 1988. v + 90pp. (11 340902 8).
104. Detention at the police station under the Police and Criminal Evidence Act 1984. David Brown. 1989. v + 76pp. (0 11 340908 7).
105. Changes in rape offences and sentencing. Charles Lloyd and Roy Walmsley. 1989. vi + 53pp. (0 11 340910 9).
106. Concerns about rape. Lorna J. F. Smith. 1989. v + 48pp. (0 11 340911 7).
107. Domestic violence. Lorna J. F. Smith. 1989. vii + 132pp. (0 11 340925 7).
108. Drinking and disorder: a study of non-metropolitan violence. Mary Tuck. 1989. v + 111pp. (0 11 340926 5).

ALSO

Designing out crime. R. V. G. Clarke and P. Mayhew (editors). 1980. vii + 186pp. (22 340732 7).
(This book collects, with an introduction, studies that were originally published in HORS 34, 47, 49, 55, 62 and 63 and which are illustrative of the situational approach to crime prevention.)
Policing today. Kevin Heal, Roger Tarling and John Burrows (editors). 1985. v + 181pp. (11 340800 5).
(This book brings together twelve separate studies on police matters produced during the last few years by the Unit. The collection records some relatively little known contributions to the debate on policing.)
Managing criminal justice: a collection of papers. David Moxon (editor). 1985. vi + 222pp. (11 340811 0).
(This book brings together a number of studies bearing on the management of the criminal justice system. It includes papers by social scientists and operational researchers working within the Research and Planning Unit, and academic researchers who have studied particular aspects of the criminal process.)
Situational crime prevention: from theory into practice. Kevin Heal and Gloria Laycock (editors). 1986. vii + 166pp. (11 340826 9).
(Following the publication of *Designing Out Crime,* further research has been completed on the theoretical background to crime prevention. In drawing this work together this book sets down some of the theoretical concerns and discusses the emerging practical issues. It includes contributions by Unit staff as well as academics from this country and abroad.)
Communities and crime reduction. Tim Hope and Margaret Shaw (editors). 1988. vii + 311pp. (11 340892 7).
(The central theme of this book is the possibility of preventing crime by building upon the resources of local communities and active citizens. The specially commissioned chapters, by distinguished international authors, review contemporary research and policy on community crime prevention.)
New directions in police training. Peter Southgate (editor). 1988. xi + 256pp. (11 340889 7).
Training is central to the development of the police role, and particular thought and effort now go into making it more responsive to current needs—in order to produce police

*Out of Print.

officers who are both effective and sensitive in their dealings with the public. This book illustrates some of the thinking and research behind these developments.)

The above HMSO publications can be purchased from Government Bookshops or through booksellers.

The following Home Office research publications are available on request from the Home Office Research and Planning Unit, 50 Queen Anne's Gate, London, SW1H 9AT.

Research Unit Papers (RUP)

1. Uniformed police work and management technology. J. M. Hough. 1980.
2. Supplementary information on sexual offences and sentencing. Roy Walmsley and Karen White. 1980.
3. Board of Visitor adjudications. David Smith, Claire Austin and John Ditchfield. 1981.
4. Day centres and probations. Suzan Fairhead, with the assistance of J. Wilkinson-Grey. 1981.

Research and Planning Unit Papers (RPUP)

5. Ethnic minorities and complaints against the police. Philip Stevens and Carole Willis. 1982.
6. *Crime and public housing. Mike Hough and Pat Mayhew (editors). 1982.
7. *Abstracts of race relations research. George Mair and Philip Stevens (editors). 1982.
8. Police probationer training in race relations. Peter Southgate. 1982.
9. *The police response to calls from the public. Paul Ekblom and Kevin Heal. 1982.
10. City centre crime: a situational approach to prevention. Malcolm Ramsay. 1982.
11. Burglary in schools: the prospects for prevention. Tim Hope. 1982.
12. *Fine enforcement. Paul Softley and David Moxon. 1982.
13. Vietnamese refugees. Peter Jones. 1982.
14. Community resources for victims of crime. Karen Williams. 1983.
15. The use, effectiveness and impact of police stop and search powers. Carole Willis. 1983.
16. Acquittal rates. Sid Butler. 1983.
17. Criminal justice comparisons: the case of Scotland and England and Wales. Lorna J. F. Smith. 1983.
18. Time taken to deal with juveniles under criminal proceedings. Catherine Frankenburg and Roger Tarling. 1983.
19. Civilian review of complaints against the police: a survey of the United States literature. David C. Brown. 1983.
20. Police action on motoring offences. David Riley. 1983.
21. *Diverting drunks from the criminal justice system. Sue Kingsley and George Mair. 1983.
22. The staff resource implications of an independent prosecution system. Peter R. Jones. 1983.
23. Reducing the prison population: an explanatory study in Hampshire. David Smith, Bill Sheppard, George Mair and Karen Williams. 1984.
24. Criminal justice system model: magistrates' courts sub-model. Susan Rice. 1984.
25. Measures of police effectiveness and efficiency. Ian Sinclair and Clive Miller. 1984.
26. Punishment practice by prison Boards of Visitors. Susan Iles, Adrienne Connors, Chris May and Joy Mott. 1984.
27. *Reparation, conciliation and mediation. Tony Marshall. 1984.
28. Magistrates' domestic courts: new perspectives. Tony Marshall (editor). 1984.
29. Racism awareness training for the police. Peter Southgate. 1984.
30. Community constables: a study of policing initiative. David Brown and Susan Iles. 1985.
31. Recruiting volunteers. Hilary Jackson. 1985.
32. Juvenile sentencing: is there a tariff? David Moxon, Peter Jones and Roger Tarling. 1985.

---

*Out of Print.

33. Bring people together: mediation and reparation projects in Great Britain. Tony Marshall and Martin Walpole. 1985.
34. Remands in the absence of the accused. Chris May. 1985.
35. Modelling the criminal justice system. Patricia M. Morgan. 1986.
36. The criminal justice system model: the flow model. Hugh Pullinger. 1986.
37. Burglary: police actions and victims' views. John Burrows. 1986.
38. Unlocking community resources: four experimental government small grant schemes. Hilary Jackson. 1986.
39. The cost of discriminating: a review of the literature. Shirley Dex. 1986.
40. Waiting for Crown Court trial: the remand population. Rachel Pearce. 1987.
41. Children's evidence: the need for corroboration. Carole Hedderman. 1987.
42. A preliminary study of victim offender mediation and reparation schemes in England and Wales. Gwynn Davis, Jacky Boucherat and David Watson. 1987.
43. Explaining fear of crime: evidence from the 1984 British Crime Survey. Michael Maxfield. 198.
44. Judgements of crime seriousness: evidence from the 1984 British Crime Survey. Ken Pease. 1988.
45. Waiting time on the day in magistrates' courts: a review of case listing practices. David Moxon and Roger Tarling (editors). 1988.
46. Bail and probation work: the ILPS temporary bail action project. George Mair. 1988.
47. Police work and manpower allocation. Roger Tarling. 1988.
48. Computers in the courtroom. Carole Hedderman. 1988.
49. Data interchange between magistrates' courts and other agencies. Carol Hedderman. 1988.
50. Bail and probation work II: the use of London probation/bail hostels for bailees. Helen Lewis and George Mair 1989.
51. The role and function of police community liaison officers. Susan V. Phillips and Raymond Cochrane. 1989.
52. Insuring against burglary losses. Helen Lewis. 1989.

Research Bulletin

The Research Bulletin is published twice a year and consists mainly of short articles relating to projects which are part of the Home Office Research and Planning Unit's research programme.

*Out of Print.

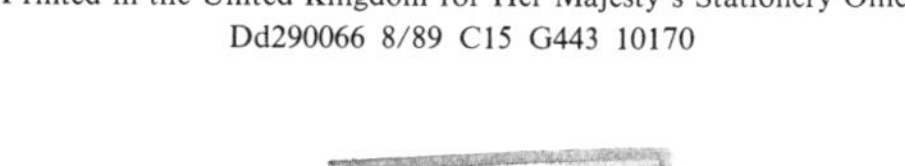
Printed in the United Kingdom for Her Majesty's Stationery Office
Dd290066 8/89 C15 G443 10170